Zones of Re-membering

AF758340

Consciousness Literature & the Arts 28

General Editor:
Daniel Meyer-Dinkgräfe

Editorial Board:
Anna Bonshek, Per Brask, John Danvers,
William S. Haney II, Amy Ione,
Michael Mangan, Arthur Versluis,
Christopher Webster, Ralph Yarrow

Zones of Re-membering
Time, Memory, and (un)Consciousness

Don Gifford

Edited by Donald E. Morse

Rodopi

Amsterdam - New York, NY 2011

Cover illustration: Solitary Cedar (Magányos Cédrus) Csontváry Museum, Pécs, Hungary (on loan from the Hungrian National Gallery). Reproduced with the permission of Janus Pannonius Museum, Pécs.

Cover design:
Aart Jan Bergshoeff

The paper on which this book is printed meets the requirements of "ISO 9706:1994, Information and documentation - Paper for documents - Requirements for permanence".

ISBN: 978-90-420-3259-0
ISSN: 1573-2193
E-Book ISBN: 978-90-420-3260-6
E-book ISSN: 1879-6044
© Editions Rodopi B.V., Amsterdam - New York, NY 2011
Printed in the Netherlands

Contents

2. Time, Memory, and the Unconscious

Preface

The late Don Gifford was Professor of English and Class of 1956 Professor of American Studies Emeritus at Williams College in Williamstown, Massachusetts. Best known among his many publications are *The Farther Shore: A Natural History of Perception* (1990) and *Ulysses Annotated: Notes for James Joyce's Ulysses* (1974, revised edition 1988). *Zones of Re-membering* was originally given by Professor Gifford as six lectures under the general title, "Zones of Memory" for the Berkshire Institute for Lifetime Learning in Massachusetts during September-October 1995. The lectures were revised slightly before Professor Gifford's death but remain substantially as he gave them. Part of their impact lies in the speaking voice meditating on this formidable subject of art, memory, time, and consciousness. Of the other two lectures, the first, "The Imitation of Dream in Literature" appeared as a chapter in *The Delegated Intellect: Emersonian Essays on Literature, Science, and Art in Honor of Don Gifford* (1995), which I edited. The second "A Chip on His Shoulder or One for the James Joyce Centennial" was published in the *Hungarian Journal of English and American Studies* (3.2 [1995]). This essay was one of several contributions Gifford made to that journal over a period of years. I am grateful to Peter Lang, Publishers and the *Hungarian Journal of English and American Studies* and its then editor, Professor Zoltán Abádi-Nagy for permission to reprint these lecture-essays.

As editor, I have not changed the text beyond correcting proofing errors that inevitably creep into such an extensive project and assigning some clarifying titles to sections. While I have made every effort to complete Gifford's extensive annotations, they must necessarily remain unfinished. My effort to complete the citations has been greatly aided by several librarians at the State University of New York College at Brockport, especially by Robert Gilliam, head of Inter-Library Loan. I also wish to thank Professors Declan Kiberd of University College Dublin and Steven Fix of Williams College for their encouragement and assistance. Finally, my deepest thanks go to Vera Benczik who created the index and to Szilárd Orosz who did the formatting.

Don Gifford was a formidable lecturer and a great teacher in the tradition of Ralph Waldo Emerson in that he provoked his listeners into learning. His own considerable erudition was continuously pressed into the service of illuminating ideas. These eight lectures demonstrate his extensive interests and study, his ability to synthesize reflections, to assimilate a wide-range of knowledge and to link perceptions in often highly original ways—all expressed clearly, forthrightly, and entertainingly. In all he did he took his task, his teaching assignment, seriously, but not himself. So rather than pronounce, he speculates; rather than pontificate, he muses, and we, his listeners and readers learn to speculate and muse along with him on the function and importance of mind, memory, time, and art.

It has been a great privilege to edit these papers—to hear again that voice pondering and questioning, always with good humor and always with utmost seriousness. As so often, I am once again in the debt of Robert Adolph of York University, Toronto who read and made extensive suggestions about the manuscript and contributed the splendid introduction to this book.

Donald E. Morse
Debrecen, Hungary

Copyright Acknowledgements

The editor and publisher gratefully acknowledge permission to reprint from the following:

> Chapter 7 from *The Delegated Intellect: Emersonian Essays on Literature, Science, and Art in Honor of Don Gifford* edited by Donald E Morse. Peter Lang, New York, 1995, pages 265-83. Used by permission of the editor and Peter Lang Publishers. Copyright 1995.
>
> Chapter 8 from *HJEAS: The Hungarian Journal of English and American Studies* 3.2 (1995): pages 161-76. Reprinted by permission of the editor, Zoltán Abádi-Nagy and the University of Debrecen Press. Copyright 1995 the University of Debrecen Press

Every reasonable effort has been made by the editor to trace the owners of copyrighted materials in this book, but in some instances this has proven impossible. The author and publisher will be glad to receive information leading to more complete acknowledgements in subsequent printings of the book, and in the meantime extend their apologies for any omissions. .

Introduction: Assaulting "Newton's Sleep"

Robert Adolph

Although Don Gifford plunges head first from the start of *Zones of Re-membering: Time, Memory, and (un)Consciousness* into some of the classic problems of philosophical inquiry, readers will not emerge with solutions to the great issues. His goal is not to solve the mind-body problem, or define the nature of consciousness and time, or show us how the brain works—even if all of these issues lurk in the background throughout his work. His first aim, I think, is to show us how no single, blinkered explanation can account for the depth and complexity of human experience, and in particular its grounding in Memory. In the spirit of Ralph Waldo Emerson, Gifford regards literature as a better guide to how we remember and experience the world than philosophy and science—though he knows a great deal about philosophy and science, and when it suits his purposes uses philosophical and scientific notions to great advantage.

Zones of Re-membering: Time, Memory, and (un)Consciousness, thus, an assault, in the Romantic—more specifically Emersonian—tradition on what William Blake called "Newton's sleep," the belief, stemming primarily from the Enlightenment, in scientific objectivity as the only path to knowledge. Gifford is hardly the first to take this position. His Romanticism, however, though rooted in the likes of Emerson, Henry David Thoreau, and many others, has been knowledgably and suggestively brought up to date by him for our times.

Gifford's focus in all these lectures and articles, and his entrée into his topic, is on the idea of Memory. One must begin, of course, with a definition. Like the American judge who, when asked to define pornography, replied that he couldn't define it but knew what it was when he saw it, Gifford begins with St. Augustine's classic exposition and admission in the *Confessions* (referring to his closely related discussion of time) that "I know what it is if no one asks me what it is; but if I want to explain it to someone . . . I find that I do not know"

(Augustine 1963: 267). Not abstract formulations but only poetic comparisons of Memory to other things (of which Gifford supplies a wealth of gorgeous examples) can begin to describe its richness. John Locke's reduction of the virtually infinite complexities of Memory to the passive reception of sense impression on the brain, conceived as a *tabula rasa* or blank sheet, to be arranged by "hosts of busy, anonymous archivists," is Gifford's first target. Much more of his fire is directed at the current inheritors of John Locke, for whom the mind is nothing more than a glorified computer, simply receiving and arranging data. Then there is the Freudian version of the Lockean hypothesis, that everything in Memory is meaningful and retrievable, at least to another set of anonymous archivists versed in the science of psychoanalysis. Then we come to the mechanistic theories of modern neurosurgeons: that it's all just nervous neurons and snapping synapses also rejected by Gifford on the grounds that memory is so radically un-machine-like. Memory—like experience generally—does not move, as Stephen Hawking would have it, like an arrow, but (in one of Gifford's favorite images) like a river, moving erratically backwards and forwards, filled with ebbs and flows, rapids and shallows.

A major theme of this book is how consciousness is a function of Memory. The ancient world, as Gifford tells us in his second lecture, recognized that "life without Memory is no life at all." This insight was expressed in the idea of the Muses, with Memory (Mnemosyne) as the mother of the rest of knowledge (Clio for History, Urania for Astronomy, and so on) and therefore of consciousness in general. Gifford combines his musings on the Muses (pun intended) with a discussion of how the so-called Aboriginals of Australia have preserved their traditions and civilization in their now disappearing "chant poems" or "song-lines." These ancient art-works are both utilitarian and mythic and still in use and, like the Muses, the source of all aspects of consciousness. Gifford compares current efforts in schools to work out a "core curriculum" as analogous to finding our own "song-lines."

Gifford then goes on in the third section to consider the relation between Memory and language. Language began not as strictly utilitarian (what Gifford calls its "pass-the-ketchup" function) but as mnemonic devices, as it still so often is with children. Gifford believes that the earliest literary works are products of such memorized oral traditions, like the Homeric epics and *Beowulf*. In classical times, as students of Rhetoric know, there was the widely practiced Art of

Memory, enduring into the Renaissance and merging with the Hermetic traditions. With the advent of printed books the Art of Memory lost its usefulness, replaced by silent solitary reading. Eventually the spread of print systematic records—indices, card catalogues and the like—replaced orality and mnemonic devices, and in turn these are now rapidly giving way to the computer. The Enlightenment and succeeding generations, recording culture and language in encyclopedias and dictionaries and concordances and computers, with their basis in scientific objectivity, though obviously useful, had the effect of sharply limiting the role of untutored consciousness.

In chapter four Gifford again takes on the idea that language should only serve "pass-the-ketchup" utilitarianism. Most of language, he argues, is not *acquired* but *made up*. Children half-invent language, immersing themselves in it rather than "acquiring" it by rote-learning the rules for utilitarian purposes. Language, meaning, and Memory all work together at the same time. Memory and language, however, never really accurate or stable appear as narrative. These narratives form our entire culture, or *paideia*. Conservatives like T. S. Eliot and Allen Bloom lament today the passing away of these narratives, without realizing that narratives are always changing and being re-interpreted, and their passing has always been lamented. Memory and narratives and languages themselves are always elusive, never stable. Little of them are of the "pass-the-ketchup" variety.

Chapter five considers Memory in its widest context, History. After a discussion on the coincidence of Passover and Easter, Gifford contrasts two principal ideas of historical Memory. The first is a "centralized," goal-oriented one "organized around a central core or cause, either secular or divine." Examples of such progressions are Manifest Destiny, the Course of Empire, the Triumph of the Proletariat, and Eden/Crucifixion/Apocalypse/Last Judgment. The second is a non-goal oriented idea of History which Gifford, borrowing from Benoit Mandelbrot's non-Euclidean geometry, labels "fractal" History, or to put it simply, one damn thing after another. Its great example is Charles Darwin's theory of evolution. We would like to live in a centralized world, but after Darwin we know we live in a fractal one, determined by chance. Gifford rings many changes on centralist attempts to enforce their views on others, resulting in massacres, wars, and other horrors, in which not only people but the records of their cultural Memory were targeted, as, for example, in

Adolph Hitler's attempts not only to destroy the Jews but all the artifacts of their civilization.

Chapter six explores, first, the relationship between Memory and the Self and, second, how Art is both a way of knowing and of "nourishing memory and consciousness." How crucial Memory is for the self becomes apparent in Alzheimer's disease, when in the absence of Memory the self disintegrates. When Memory is unbearable, as with many holocaust survivors, the self is profoundly affected. Nevertheless, some people in harrowing circumstances both in fiction, as in Daniel Defoe's version of Robinson Crusoe, and in real life, as survivors of solitary confinement, such as the imprisoned Hungarian intellectual Edith Bone, can survive if able to draw on reserves of Memory.

Our individual and collective Memory is stored in the arts. Art, as the Romantics knew so well, is as valid a way of knowing as science. Art does this by enriching Memory. Art-knowledge, of course, has nothing in common with the processes of computers. Through art the inhabitants of stressful places like Theresienstadt and besieged Sarajevo were restored.

Chapter seven, though not part of the initial lecture series, continues its explorations of Memory. Gifford is concerned here with a perennial literary genre, the Dream Vision. Dreams are another form of Memory. Their representation has always been problematic, mostly because they resist the realistic or "mimetic" forms of representation traditionally available to narrators, dramatists, and poets. Writers like William Meredith, Henry James, and Dorothy Richardson have tried to get around this problem by resorting to "stream of consciousness" and similar techniques. Such methods, however, usually end up as verbal representations of dreams, rather than following the non-verbal contours of the dreams themselves. Another problem is that dreams have always been interpreted as something much more than the materials of ordinary mimetic narration. They seem to shape and contain Memory at its deepest levels. In the Bible they have been seen as omens or predictions, in Freudian psychoanalysis as expressions of repressed unconscious desires, and in current psychological theory as the mind maintaining mental health.

For Gifford, the most important and successful traditional representation of dreams in literature are the medieval Dream Visions. These are not so much attempts to record actual dreams, but literary structures representing complex, significant cultural memories. Major medieval examples include *The Divine Comedy*, *The Romance of the*

Rose, and *Piers Plowman*. All are allegorical, but the allegory is never obvious. Later practitioners of the Dream Vision include John Bunyan, Lewis Carroll in *Alice in Wonderland* and *Through the Looking Glass*, and Nathaniel Hawthorne. And more indirectly, there is Shakespeare in *A Midsummer Nights Dream*, Samuel Taylor Coleridge in *Kubla Khan*, Emily Bronte in the Gothic opening of *Wuthering Heights*, Herman Melville in the Try-Works chapter in *Moby-Dick*, and, of course, August Strindberg in *The Dream Play* and *Ghost Sonata*.

In *Ulysses*, perhaps most notably in the Circe episode, James Joyce combines the dream vision with fictional narration so that it is impossible to separate the two. Finally, here and even more so in *Finnegans Wake* Joyce at last solved the problem of representing dreams as they actually happen, not as they were narrated later. In so doing Joyce created an entirely new expression and representation of Memory.

Gifford now goes on, in chapter eight, to consider how Memory operates in his favorite author, James Joyce. A real-life trip to Ireland reveals that the present-day Irish themselves do not think much of Joyce because he got so many details about the country wrong. In the west of Ireland Joyce is disliked simply because he came from Dublin. And in that city, too, Joyce was unpopular for his criticism of the Irish, especially in *Dubliners* and *A Portrait of the Artist as a Young Man*, in which the city is portrayed as a place of spiritual paralysis and "simony," the selling-off of any spiritual values. Assuming that Joyce's writings, especially the *Portrait*, are autobiographical, such Irish critics see in Stephen Dedalus an inaccurate, self-serving portrait of Joyce himself carrying around a very large chip on his shoulder against Ireland, its culture, and its people. What happened to Joyce's Memory?

Gifford insists that Joyce's critics have misunderstood him. In the *Portrait*, Joyce was showing his alter ego, Stephen, as a young aesthete and gently mocking him as much as his compatriots. Under the influence of Hendrik Ibsen, Joyce, like many of his Irish and Anglo-Irish predecessors such as Jonathan Swift, G. B. Shaw, and W. B. Yeats, moved in *Dubliners* from an art ruled by aestheticism to one concerned with ethical issues and problems. These issues and problems are worked out in *Ulysses*, most notably in the Circe, Cyclops, Hades, and Sirens sections. Gifford's analyses are too complex to summarize here, but in the end his hero Leopold Bloom, though more or less a Jew, redeems himself through acts of Christian love, as defined in terms of Augustinian *caritas*, and by so doing points the way to universal ethical redemption. The many layers of Joyce's incredible art, as Gifford says, bring out "the subtleties, equivocations and the subterranean complexities which dominate the average man and comprise his life." In *Ulysses*, in other words, every aspect of Memory is presented, and Joyce has realized his project, announced famously in the *Portrait*, to present "the reality of experience and to forge in the smithy of my soul the uncreated conscience of my race" (Joyce 1976: 154). (Presumably, Joyce means both Ireland and humanity by "my race.") *Ulysses* is to Ireland, and human Memory generally, what *Don Quixote* is to Spain and human Memory. "And such a book," says Gifford, quoting the critic John Eglinton, "need not insult . . . the cause of Irish nationality any more than Cervantes laughed real chivalry away."

For Gifford, as we see in the last two lectures, the profoundest explorer of human consciousness, time, and memory is James Joyce. Indeed in their range of reference, wit, and humanity the spirit of Joyce permeates all the lectures collected in this book.

I

Time, Memory, and Consciousness

Chapter 1
Zones of Re-membering

In *The Confessions* St. Augustine asks "What then is time?" and answers himself: "I know what it is if no one asks me what it is; but if I want to explain it to someone who has asked me, I find that I do not know" (Augustine 1963: XI.14, 267). William James adopted a similar rhetorical strategy in response to the question "What is consciousness?"—"We know as long as no one asks us to define it." And he remarks that it would be helpful "[i]f we could say in English 'it thinks,' as we say 'it rains' or 'it blows'" (William James 1981: I:220). And so with memory: as long as no one asks me, I know what it is. It is, as Oliver Sacks remarks (quoting Frederic Bartlett) not the noun "memory" but "the dynamic process of 'remembering'" (Sacks 1995: 172-73), not the noun but always the verb, not an entity, but always a process. And yet, in my experience, it remains a habitation to which I am so habituated that definition eludes me, and I search constantly for satisfying metaphors for the dynamics of its space/time.

But I am pleased that the same rhetorical strategy works for each of the big three: time, consciousness, and memory, an ever-shifting but unbreakable triangle. I say unbreakable because if one side of the triangle is removed, the other two sides dissolve and disappear. Not that consciousness, memory, and time are synonymous but that they are interwoven in the triangular web of our experience to the point where they are frequently all but interchangeable: "I'm talking about memory."—"Oh, you mean consciousness."—"No, I was actually thinking about time." In much of the reading I've been doing on these subjects in the past several years, the three terms seem rhetorically interchangeable, as in the three questions of my opening paragraph: What then is time? What then is consciousness? What then is memory?

St Augustine remarked: "The fields, the caves, the dens of memory cannot be counted." And in another context, he expanded on that:

> How great . . . is this force of memory, how exceedingly great! It is like a vast and boundless subterranean shrine. Who has ever reached the bottom of it? Yet this is a faculty of my mind and belongs to my nature; nor can I myself

> grasp all that I am. Therefore, the mind is not large enough to contain itself. But where can that uncontained part of it be? Is it outside itself and not inside? In that case, how can it fail to contain itself? (Augustine 1963: X.8, 219)

That triad of questions remains unanswered after sixteen centuries, and here I am (we are) still awash in memory, in its multi-channeled, riverine behavior, its rapids and deep pools, its eddies and backwaters; the flow of its current, the odd ways it has of floating things to the surface or carrying them under—sometimes with the feeling that the flotations and disappearances are from the body and the surround as well as from the "in here" of the brain case. Aqueous and especially riverine metaphors have come to dominate my summer thinking about memory.

But a host of other metaphors waits in the wings: there are St. Augustine's fields and caves and dens, his sanctuary. There are houses, palaces, theatres, churches and chapels as well as shady groves, forests and jungles. Medieval and Renaissance psychologies divided the mind up into innate (inborn) faculties such as feeling, perception, understanding, memory, etc.—each an entity unto itself. And that faculty psychology brought with it metaphors of containers: chests, storage bins, closets, hampers, cupboards; and those remain to haunt our common sense when we imagine memory as a container and think of it as limited in its capacities. A close friend once surprised me by remarking that he had long ago decided that he had to choose between an interest in classical music and an interest in jazz because he would not have the memory storage-room for both. I remember being surprised by his remark, but I do not remember the why of my surprise since, in the quirky way memory has of up-dating things, I have a now-reason for my surprise which I do not think was available to me then.

In the *Essay on Human Understanding* (1690) John Locke (1632-1704) attempted to sweep those innate faculties and their containers away in favor of a growth model: the mind conceived as a *tabula rasa*, a smooth tablet on which the stylus of sense impressions traces images that combine and build into memory and understanding—all based on the empirical groundwork of sensory experience, and resulting in what Locke called "determined ideas": "Some immediate object of the mind, which it perceives and has before it, distinct from the sound it uses as a sign of it" (Locke 1986: 1757).

At a stroke Locke had swept away "objects of faith" and other abstractions in favor of empirical solidities. Within a generation Locke's "determined ideas" had been over-simplified and reduced to the status of "facts"—leading to what Hugh Kenner has called "this tyranny of Fact" in the eighteenth and nineteenth centuries (Kenner 1968: 61), a concept of human memory and understanding machine-tooled to accommodate the machinery of the Newtonian universe. This concept in turn encouraged the view that memory was a stored collection of fixed images of these "immediate objects of the mind"—reductively, a stored and retrievable collection of "facts."

This assumption that memory was a vast accumulation of fixities encouraged mechanical images of memory as portfolios of prints and written records, an archive staffed by busy, anonymous archivists, and still to come: card files, photograph albums, collections of recordings and film clips, a library of variously stored and ideally retrievable sights, sounds, smells, tastes, plus physical impressions of touch, of kinesthetic states and movements, and all kinds of mental images. Recently and climactically this mechanization of memory and its brain (or the brain and its memory) has been imaged as and likened to the computer.

I want to pause here for a reality-check: we all know enough about the fallibilities and uncertainties of our memories to give us cause for wonder about how fixed and readily available to recall those supposedly vast quantities of representations (or fixed images) are. When we are assured by the memory-optimists that all our personal experiences are stored somewhere "in there," I immediately sanitize that mythical zone "in there" with quotation marks. Puzzled by that optimistic assurance, we may well wonder and pause to contemplate Jorge Luis Borges's story "Funes the Memorious" in which Ireneo Funes, when he sets himself the task of remembering a day, can readily accomplish that task but with such an incredible wealth and specificity of detail that it will take him a full day to do so: a staggering, not to say frightening price to pay for the ambiguous gift of total representational recall. That "price" is spelled out in the story; imaged in the paralysis that has immobilized Ireneo and rendered his "perception and his memory . . . infallible" (Borges 1964: 63). And Ireneo's friend, the narrator of the story, suspects that part of the price has been that Ireneo is "not very capable of thought" because "[t]o think is to forget differences, generalize, make abstractions. In the teeming world of Funes, there were only details, almost immediate in their presence" (Borges 1964: 66).

This perception effectively paralyzes the narrator as well when he is at his friend's bedside: "I was benumbed by the fear of multiplying useless gestures" because every gesture "would persist in his implacable memory" (Borges 1964: 66). Borges's story is, to all intents and purposes, a surreal meditation on the folly of the memory-optimists and their belief in total storage and the possibility of total recall. In Borges' story this folly has the sobering consequence of paralysis, an unthinking paralysis, and subsequently death "of congestion of the lungs"—all too probable for a person immobilized by physical paralysis and without artificial ventilation as Funes is; and all too symbolic: to breathe in is to perceive, and to perceive for Funes is to remember in such fantastic detail that the end result can be nothing other than terminal congestion.

Freud was a staunch believer in the total-storage, it's all-in-there-somewhere theory of memory, but with a proviso: during his self-analysis, he taught himself to remember his dreams in minute detail. Eventually he reports that he had to unteach himself or rather that he had to teach himself to forget lest, he said, he became like the Sphinx, "drowning in the sands of the Nile." We might remind ourselves that in the Greek myth of Oedipus the Sphinx was a propounder of riddles who had reduced the city of Thebes to a waste land and who destroyed those who sought to help the city but who could not answer her riddle. On the way to fulfillment of his fated role in the supremely dysfunctional family, Oedipus confounds the Sphinx by answering her riddle and thus he becomes King of Thebes and his mother's husband. But Freud had in mind not Oedipus's sphinx but the great Sphinx at Giza in Egypt on the edge of the Sea of Sand. By the beginning of the nineteenth century that sphinx was buried in sand up to its chin though by Freud's time it had been partially exhumed.

Contemporary neurophysiology and neuropsychology call the total-storage, total-retention theory of memory seriously in doubt. If for centuries we have regarded memory as an entity, the neurosciences are now suggesting that it is an elaborate pattern of processes rather than an entity, and that it is not a matter of storage and retrieval of "representations" (or fixed images) of the past but a matter of procedures "created . . . developed over time" (Rosenfield 1988: 128, 127). The new theory is that memories are not "fixed memory traces" but "'interpretations' of previous impressions in terms of present circumstances" (Rosenfield 1988: 75). In other words, memory is always present tense, the reverberations of the past brought to bear on, to contextualize, inform, and enrich experience in the here and now. Neuroscientists liken this to a continuous process of mapping and remapping as we find our way in the dailyness of consciousness. They also liken this process to the free-flowing, improvisatory nature of language as we use it from moment to moment, not always sure of what we're going to say next until we hear ourselves saying it. Some of memory's procedural centers seem to be localized in all normal human brains, but other centers and maps vary from person to person; some neuroscientists even suspect that they vary widely from person to person. In sum: "It is not different representations [fixed images] but different procedures that incorporate our varied experiences, that the varieties of art reflect" (Rosenfield 1988: 143). If we invert that perspective we can suggest that the extraordinary variety of images produced in the arts and the corresponding range of our abilities to respond are a measure of the fantastic vocabularies of procedures of which we collectively and individually are capable.

To repeat: the neuro-sciences are exploring the proposition that memory in all its varieties is mapping and remapping in the present tense of consciousness; impressions from the past are being reworked, modified, recombined in the service of present moods, interests, needs, purposes. The new emphasis is not on opportunism and dishonesty (those are always with us no matter what our model of memory), but rather on the nature of memory as process—its so-called "survival function" in the here and now of experience.

St. Augustine had an intuitive glimpse of this present-tense nature of consciousness:

> It is now, however, perfectly clear that neither the future nor the past are in existence, and that it is incorrect to say that there are three times—past, present, and future. Though one might perhaps say: "There are three times—a present of things past, a present of things present, and a present of things future." For these three do exist in the mind, and I do not see them anywhere else: the present time of things past is memory; the present time of things present is sight; the present time of things future is expectation. (Augustine 1963: XI: 20, 273)

But even as the modern neuro-sciences explore these new/old concepts of the flexibility of mnemonic processes, we are variously besieged by the vigorous presence of the old order: the belief in fixed images or representations in the permanent storage of a potentially total recall. The recent fascination with "recovered memory" offers a case in point. Recovered memory is much in the news and all over the place in talk show psychology: it is the sudden (and to the memorist very convincing) memory of an event of abuse or violence in a previously hidden past. The "recovery" pivots on the assumption that the relevant images are fixed and permanently stored "in there." As for the crimes and abuses thus recalled, there are questions, both psychological and legal. In legal terms the recovered memory is suspect if unsupported by corroborating evidence. It is so suspect that some judges (beginning with one in New Hampshire) have barred such uncorroborated memory-evidence from court on the basis that when there is so much controversy about the validity of these memories in the scientific community, there is no firm legal basis on which to proceed. The controversy in the scientific community is in part spurred by the emerging concepts of memory as procedures in process—constantly re-mapping, rewriting the past into the present tense of experience. And the controversy is exacerbated by the suspicion that wishful thinking or inept or irresponsible therapy may plant and encourage memories by suggestion rather than by recovery. For example: it is a textbook assumption that multiple personality disorder is caused by sexual abuse in childhood. If the therapist makes that diagnosis and accepts that assumption, he may, in effect, say to the patient: "it's in there somewhere. Let's go find it." This suspicion is, in part, supported by the clinical knowledge that while patients under hypnosis can be helped to recall previously forgotten events, they can also be induced to become highly creative inventors of fictions.

And when it becomes fashionable (as it is nowadays) to portray oneself as the victim at least of a dysfunctional family if not of outright crime, many of us can make do by sorting and recombining memory traces toward that self-dramatization, that purpose. One symptom of just how fashionable the new victimology has become is provided by the celebrated pop-therapist who casually estimates that between 90 and 95% of our fellow countrymen are the secretly wounded victims of dysfunctional families (see Bradshaw 1990). Another unfortunate symptom is provided by the *Diagnostic and Statistical Manual of Mental Disorders* (June 1994) which identifies more than three hundred "mental disorders" and so inter-confuses science with social and political values that something on the order of "half of all Americans [are suffering] from psychiatric disorders" of some sort or other! I find myself wondering where the "science" is in the fog of questions and answers that has resulted from this labor of the American Psychiatric Association.

But the I-am-a-victim syndrome will, I hope, gradually succumb to the evanescence of fashion. The metaphor that likens the brain and its memory to the computer (the computer syndrome) may be much harder to dislodge. Two avenues of approach, two garden paths lead toward the assumption that brain/memory is or is like a computer, and I want to consider and reconsider them at some length.

First is the assumption that cognitive actions for a human being are what is called "information processing." A generation ago the words "information" and "knowledge" were fairly close synonyms: *information* meant something that had been considered, shaped, given form and *knowledge* meant something that had been established by an act of meaning. But in the phrase "information processing" information is reduced to raw data, what the mind passes over as bits of news, factlets from here and there—as yet to be formed into "information" in the older sense of the term. In the jargon of the computer programmer, information is a "non-accidental signal used as an input to a computer." I love that term "non-accidental" because the percepts that come my way in the to and fro of the day's flux are already (and inevitably) inflected by the accidents of *affect* and *evaluation*, by the resonances of feeling and judgmental response. The "information" the computer processes is value-neutral, simply itself, but, as Auden said, "because they are endowed with memory, for humans nothing is ever simply itself." And the computer doesn't have a body to pleasure and encumber it, but we do; and as Israel Rosenfield reminds us: "The body . . . is the brain's absolute frame of reference," and, given that frame of reference, memory is radically "different from any image-retaining capability in a machine" (Rosenfield 1992: 45).

And there are other pleasures and encumbrances: the "mind of this flesh" carries about with it "the rough and perpetually changing draft of [its] . . . autobiography" (Bruner 33), and each new percept, each quantum of experience must thread the writing workshop that is perpetually engaged with that draft. Henry James, attempting to blow away the tired commonplace that one should write from experience—as presumably one should in the autobiography workshop—defines experience in such a way as to challenge the commonplace:

> Experience is never limited, and it is never complete; it is an immense sensibility, a kind of huge spider-web of the finest silken threads suspended in the chamber of consciousness, and catching every air-borne particle in its tissue. It is the very atmosphere of the mind; and when the mind is imaginative . . . it takes to itself the faintest hints of life, it converts the very pulses of the air into revelations. (Henry James 1884: 12).

For example: as I wrote a draft of this passage in my Journal, I was on Nantucket (25 July 1995). I copied Bruner's phrase and caught a flicker of that well-known passage from Henry James. At the same time I was aware of the wind in the pines outside, the stir of the curtains at the window, the alerted skin of my forearm, and the blue hydrangea tossing in the hedge below at the edge of the lawn. Nor do the "air-borne particles" stop there. That "huge spider-web" also (usually unconsciously) monitors the proprioceptors that check the skeleton and its postures and the disposition and tone of the muscle systems and tendons in play, right down to the angle of the pen in the saddle between the thumb and crooked fingers of the right hand, the left hand there to tell the page to be still. Synchronicity. Nor does the catalogue of its in-touch-with stop even there. The four peaches ripening on the window sill are perceived as a unit of four and therefore do not have to be counted; the room behind my back was occupied but did not have to be checked; the fog was lifting; arrivals pending. As I rewrite it now, I contrast the synchronicity of that experience of moment-to-moment with the time involved in the writing or rather in spelling it out, an exercise which I rarely undertake unless I am trying to make a point (as now). But the computer would have to go through the whole spelling-out, or rather, would have to have all these bits and pieces non-accidentally input into it (whatever its faster-than-I-am speed) because the computer would be "processing information" while I'm busy making a point, making meaning; trying at one and the same time to see the experience of that 10:23 a.m. moment clearly and to explore ways of writing, speaking, making it clear to you, and then I suddenly realize that you are on the far side of Labor Day from the July morning I inhabit and that realization threatens to unload a steamer trunk full of post Labor Day associations, the clutter of the beginning of another academic year and all that goes with it. I call a halt.

But to suggest the magnitude of that "rough and perpetually changing draft of [an] autobiography," a magnitude that awaits and anticipates "every air-borne mote and particle" of experience (in contrast to the computer's *tabula rasa*) there is the Turing Game. It was invented by the English mathematician Alan M. Turing (1912-54). It is a game for three players: "One player is a machine . . . trying to pass itself off as a person. Another is a person, trying to make it clear that he *is* a person. The third is an observer, trying to decide which is which" (Kenner 1968: 121). In order for the machine to stand a competitive chance, Turing calculated that it would have to be supplied with 10^7 bits of "information"—a modest one hundred million bits and pieces of data which would provide sixty programmers with fifty years of steady work "if," Turing said, "nothing went into the wastepaper basket." (In view of the numbers to come later in this lecture, that 10^7 is probably an underestimate, but we can proceed without quibbling over the point.) Presumably those 100,000,000 bits would include a "rough and perpetually changing draft of [an] autobiography," complete with lapses of memory and misrememberings, plus the ability to make mistakes and the ability to make lucky guesses. Plus a poser: in the course of an hour's conversation the machine-player's autobiography would not only be reviewed, it would also have to undergo an extensive rewrite and how is that capacity for rewrite to be programmed into the machine-player's "mind"? The point is that each time I retell this or that phase of my life-story, I always revise and update it for the occasion. How is the machine to be instructed to do that?

As for the observer-player, he might be inclined to favor the machine-player's "humanity" if he had been educated by those brain researchers who report (*New York Times*, 30 May 1995) that the computer has given them "a way to think about information processing" in the human brain. Or, as Hugh Kenner has said: "Turing was [in 1950] perhaps not allowing for the possibility that people will grow more machine-like" (Kenner 1968: 123), or, put another way: "The computer simulates thought when thought has been defined in a computer's way" (Kenner 1968: 40).

Back to the neutral ground between the two garden paths that lead to the brain/computer metaphor, what I call the computer syndrome. If one of those paths is the "information processing" path I've been trying to characterize, the other is a less direct, less deliberate mind/computer analogue. The neuro-sciences have made fabulous brain progress in the last few decades. Thanks to new brain-scanning techniques and to advances in bio-chemistry and molecular biology, explorations of the labyrinthine potential, the all but infinite possibilities of the brain's synaptic highways and byways, the numbers invoked become so astronomical as to be beyond our grasp. In *Bright Air, Brilliant Fire* Gerald Edelman remarks that the normal cerebral cortex is composed of "about ten billion neurons" with the potential for "about one million billion connections" (that's 10^{14} connections) (Edelman 1992: 17). And he adds:

> Another way of getting a feeling for the numbers of connections in this extraordinary structure is to consider that a large match head's worth of your brain contains about a billion connections. Notice that I only mentioned counting connections. If we consider how connections might be variously combined, the number would be hyperastronomical—on the order of ten followed by millions of zeros. (There are about ten followed by eighty zeros'

worth of positively charged particles in the whole known universe!) (Edelman 1992: 17).

In sum: the human brain is "the most complicated material object in the known universe" (Edelman 1992: 23). Edelman explicitly states that we can never machine-model the brain because "the variation is too great" (Edelman 1992: 25). But those hyperastronomical numbers (my brain is bigger than the universe) are beyond imagining—just as I can't imagine the duration of the Big Bang (10^{-46} seconds)—though I wouldn't have missed the occasion for either of those numbers for the world.

I find it fascinating that almost all of the recent books on brain/mind/memory/consciousness compare the dimensions of the mind to astronomical dimensions. It is as though Bertrand Russell's (1872-1970) observation in an essay "How to Read and Understand History" (1943) were turned inside-out: "What astronomy does to enlarge the spatial habitat of the mind, history does to increase its temporal domain" (Russell 1943: 23). We cannot even imagine these new astronomies, let alone crunch such numbers by hand; only the super-computers and their jockeys can juggle and manage such overload. [Aside: I am reminded of the 16-year-old Thomasina in Tom Stoppard's *Arcadia* (1993). Before her immolation in 1817 she has entered on a sequence of calculations in her notebook which, if pursued over several lifetimes, would have resulted in fractal geometry. Today, hundreds of the computers in this college town could manage those lifetimes in a matter of minutes.] Back on track from that aside: the second garden path thus looms: if only a computer can handle these computations, then the brain must be like a computer.

But this comparison, this metaphor is as a mirror's image of a person standing in front of it—the person (the mind and its brain) can gesture and turn and walk away. The mirror (the computer) is fixed in two dimensions plus a frozen digital time, including the time to be blank and unresponsive: empty of image, lights out. The brain can count, but it does not deploy and inhabit its astronomical range by counting, and so we are back to a new twist on Russell's enlargement of the brain's space/time via the contemplation of astronomy and history.

I do not mean here to denigrate or call in question the fabulous recent achievements of the neuro-scientists but to suggest that when they have finished, wrapped it up and put the copestone in place, we will be non-the-less mystified by the present tense and phenomenology of our memories, our consciousnesses, and our experiences of time. Indeed many neuro-scientists, Edelman among them, warn repeatedly and eloquently against our accepting what Edelman calls "a machine model" of the mind (Edelman 1992: 171). He goes on:

> Mechanism now lives next to fanaticism. Societies are in the hands either of the commercially powerful but spiritually empty or . . . in the hands of fanatical zealots under the sway of unscientific myths and emotion. Perhaps when we understand and accept a scientific view of how our mind emerges in the world, a richer view of our nature and more lenient myths will serve us. (Edelman 1992: 171)

Several pages later he concludes that consciousness "will always remain richer than its [analytic] description" (Edelman 1992: 209). In short, once computers have enabled the tool-users to understand the bio-genetics of the brain and its mind, we will be, perhaps, even more richly than ever, right where we are.

But no matter how firmly I am convinced that the computer/brain metaphor is a false lead, I have to pause to put in a good word for the computers that enable us to prove it. The machine at my elbow is incredibly good at algorithms, those mechanical and recursive procedures we associate with linear logic, but it invites the rebuke that the physicist Niels Bohr (1885-1962) once delivered to one of his students: "You are not thinking, you are just being logical." But that machine also has the "perfect memory" that I don't enjoy, and it is the ideal "information processor," much better than the anonymous file clerks at work back stairs in my mind, but those faceless clerks go on their butterfingered way while up here in the available light I go about the business of *gnosis*, seeking to know who we are and where we're at.

And so back to memory: as long as no one asks me, I know what it is. It is remembering all the time whether I'm paying attention or not; whether I ask it to or not; and it's enough of an anarchist to carry on even when I ask it to change the subject or to stop singing the same tune over and over, to be still. As I've eavesdropped on it this summer in preparation for what I might say here, I have found myself more and more drawn to aqueous metaphors as I've already suggested: awash in the multi-channeled riverine behavior of remembering, fascinated by its shallows and its depths and upwellings, its eddies and backwaters, the way things sink out of sight and then surface thanks to an unexpected quirk of the undercurrent.

That riverine indeterminacy is what sends things off into the underbrush when Stephen Hawking is discussing "human memory" and "time's arrow" in *A Brief History of Time*. The time's arrow bit puzzled me until I realized that he was addressing the question of time from somewhere outside the solar system, in the realm where mathematics becomes real. From that perspective all of us (and our computers into the bargain) are encapsulated in that arrow, hell-bent in the direction "in which . . . entropy increases" (Hawking 1996: 149). Hawking says: "It is rather difficult to talk about human memory because we don't know how the brain works in detail. We do, however, know all about how computer memories work" (Hawking 1996: 151); so we'll make the what-he-calls "reasonable" assumption "that the arrow [of time] for computers is the same as that for humans" (Hawking 1996: 151). The assumption is "reasonable" because it eases the way and lets him get on with the discussion except that what ensues bears little relation to the nature of mnemonic experience and therefore to our subjective sense of time and its arrow. So when he says "just as a computer, we must remember things in the order in which entropy increases" (Hawking 1996: 151), me and my memory respond with a lively *yes, but*. Yes, we, our bodies and brains and the ecosystems we inhabit are locked into "the order in which entropy increases" by a thermo-dynamic definition called the Second Law. But the computer and its "perfect memory" remembers in a digital clock time in which there is no present tense except the command from the keyboard. (There is no counterclockwise in digital time.) But my human memory is responding to all kinds of clocks, some of them turning counterclockwise. In its riverine present tense it insists on jumbling and scrambling the order, "the direction . . . in which [entropy] disorder increases" (Hawking 1996: 151). Not that my memory isn't increasingly disorderly, but the time in which it is disorderly does not feel like an arrow flight but a river as many streamed as memory itself. Or, as Isaiah Berlin once remarked: "The present is not predictive of the future; the present is 'retrodictive' of the past." Subjectively speaking, time does not feel very much like an arrow flight, but repeatedly addresses us in the person of Emerson's "Brahma":

> If the red slayer think he slays,
>   Or if the slain think he is slain,
> They know not well the subtle ways
>   I keep and pass and turn again.
>
> Far or forgot to me is near;
>   Shadow and sunlight are the same;
> The vanished gods to me appear;

> And one to me are shame and fame. (1856/1857)
> (Emerson 1996: 1209)

In the absence of memory "time . . . [comes] to a stop" as Oliver Sacks observes of two of his variously afflicted patients in *The Man Who Mistook His Wife for a Hat* (Sacks 1985: 41). But the "go" of time from which we observe those patients is not the rainbow trajectory of an arrow, from release to gravity's return. We know its "subtle ways" better than that. But, when in Tom Stoppard's *Rosencrantz and Guildenstern Are Dead* (1967) the Player King jovially remarks to that beleaguered pair "[t]here is only one direction, and time is its only measure," we know full well what their direction is (Stoppard 1967).. They go to their deaths, as offstage in the closed world of *Hamlet* they have been doing for almost four centuries. And we may even catch an echo of the direction in which we are going in theirs except that once again this year we will pass the anniversaries of our deaths without knowing it and therefore with a different sense of the closure that so clearly is to complete the arrow flight (?) of Rosencrantz's and Guildenstern's time. I stumbled into a wonderful example of the point I'm trying to make several years ago during a seminar on T. S. Eliot's *The Waste Land* (1922). I had outlined one conventional reading of Eliot's spiritual career: from the Inferno of *The Waste Land* to the Purgatorio of *Ash Wednesday* (1928) to the Paradiso of *Four Quartets* (1942). A colleague who was sharing the seminar with me asked: what if he had committed suicide instead of *Ash Wednesday* and the conversion to Anglo-Catholicism that it celebrates? Then, would we read *The Waste Land* not as the dark night of the soul preparatory to conversion but as an utterance of terminal despair?

Science, I'm afraid, will continue to have trouble with human memory because science aspires to be predictive as well as descriptive, and memory insists on the personal spaces of its anarchy, defying description, let alone prediction: "Do you know what my memory did to me this morning?" is almost on a par with: "Do you know what my dreams did to me last night?"

Freud, committed as he was to an ironclad psychological determinism, had considerable trouble with the waywardness and vagaries of his patient's memories. And some of his recent followers have come to the conclusion that if the analyst and analysand can evolve a "narrative truth," an account of the patient's troubled past that is satisfactory in the sense that it *heals*, makes whole and "free," then that narrative truth may matter more than the "historical truth" which may remain somewhat "obscure" thanks to the intractable nature of human memory and the absence of corroborating evidence (see Donald P. Spence). So even though the past may have determined the present, that determinism can be at least deflected or even modified by the present tense of narrative truth. This fits with the emerging view among neuro-scientists that memory is insistently present tense, reshaping recollections of the past in order to inform and contextualize today's experience.

Part of the trouble lies, I suspect, in or near the concept of scientific "objectivity." When B. F. Skinner (b1904) and the behaviorists tried to reduce human behavior to the stimulus/response mechanisms that worked so well with laboratory animals, they were being *objective*. They were exposing themselves to James Thurber's quip: "All scientists are the same man," and to Vicki Herne's turn on Thurber's quip: they are striving to be "no one in particular," trying to drop out of the human equation the intractable subjectivity which *is* our experience of memory, consciousness, and time. And the present fashion of defining the brain's processes on the computational model of the computer is one more attempt at anonymous objectivity, substituting machines for lab animals, and in-put/out-put for Skinner's stimulus/response, once again inviting Alfred North Whitehead's (1861-1947) wonderful after-dinner send-up of B. F. Skinner: "Let me [hear] you account for my behavior as I sit here saying 'no black scorpion is falling on this table.'"

In *The Fragile Species* (1992) Lewis Thomas, medical scientist and essayist, traces one sequence of neuro-biological research on Intracranial Self-Stimulation (ICCS)—stimulating the so-called "pleasure centers" in the brains of laboratory animals and discovering that the animals, when put in control of the stimulus, want the "pleasure" to go on forever (Thomas 1992: 31). He reports that the neuro-biologists cleaned up their act and backed away from the term "pleasure centers" because it was too non-objectively anthropomorphic (Thomas 1992: 33-34). But Thomas insists on twitting them by continuing to use the contaminated phrase. He reports that at length they had taken their study "as far as it can be taken by skilled reductionist research" and remarks: "Personally, I like it this way" (Thomas 1992: 34). What he likes is that the laboratory stops there and leaves the subject of pleasure open to speculation about the role it plays in what he calls "The Life in the Mind." Elsewhere he suggests that the so-called "social sciences" need a new scientific method, a new way of asking questions that does not *reduce* but that expands into realms now dismissed as "anecdotal," insufficiently objective and therefore not quantifiable.

I suspect that the skilled reductionist research on the human brain that is so exciting and rewarding to the neuroscientists in our midst will lead to a similar vantage point beyond which we will need a new scientific method that can handle the intractable subjectivity, the intractable anecdotalism of the life of the mind and its memories. What that new scientific method and its questions might be, Thomas does not say, but I suspect he gives us clues in his style, his insistently contemplative and anecdotal way of seeing—as he returns again and again to explore beyond the borders of the safely objective, statistically convincing realms of "science," quoting, as he does with apparent relish, Henri Poincaré's great remark: "Science is built up with facts as a house is with stones, but a collection of facts is no more a science than a heap of stones is a house" (Quoted in Thomas 1992: 72).

Thomas's phrase "skilled reductionist research" suggests the possibility of a skilled expansionist research to match; a skill that we have yet to evolve. But it's pleasant to be able to report that William James spent considerable time at a similar cross roads one hundred years ago, denigrating what he called "[t]he notion that sensations, being the simplest things, are the first things to take up" in the study of human consciousness (James, William 1981: I:219). He advocated instead that that study and research should begin with the full complexity of the brain and its mind.

Back to memory: the film-maker Luis Buñuel in his *Memoirs* contemplates his mother's decline into senile amnesia: "You have to begin to lose your memory, if only in bits and pieces, to realize that memory is what makes our lives. Life without memory is no life at all . . . Our memory is our coherence, our reason, our feeling, even our action. Without it, we are nothing" (Buñuel 1983: 4-5).Without memory there is only contingency, no continuity, no coherence, no rhythmic interchange between the what has been, the what is, and the what is to be; or, as John Dewey (1859-1952) remarked in *Art as Experience*: "Every living experience owes its richness to what Santayana calls 'hushed reverberations'" (Dewey 1939: 18); and those reverberations, that richness, is the presence of the zones of memory.

Chapter 2
Ancient Greeks and Aboriginal Australians

Two propositions: my mind to me a computer is not, even though it may look like a computer to somebody else. When we speak of someone as "well-informed," we clearly mean "not ignorant," but the information of "information processing" does not necessarily cure ignorance.

In this lecture I want to explore two alternate routes to the phenomenology of memory: one by way of the Ancient Greeks, the other by way of the Aborigines of Australia. In both cases I am looking for the leverage of contrast, ways to bring the foreign and strange of another culture to bear on the all too familiar individual and collective memories in which we are immersed.

The Ancient Greeks
In the ancient Greek scheme of things there seem to have been three early Muses: Mnemé (remembrance), Meleté (Meditation), and Aoidé (song). The poet Hesiod who flourished in the eighth century BC elaborated the early three into nine and elevated Mnemé in the person of the goddess Mnemosyné (memory) to the position of mother of the Muses; Zeus, the all-father, stood in for nine nights as father.

Meleté, meditation
Aoidé, song.
Urania, astronomy
Melpomene, tragedy
Thalia, theatre,
Terpsichore, dance
Clio, history
Calliopé, epic poetry
Erato, lyric poetry

You'll notice from the listing of the Muses that all of them (with the possible exception of Urania) are associated with voices and song. And unlike our interiorized, secret Muses, at least eight of them and their arts involved performance, public reading, or rather public memory in recitation and improvisation. How much from memory, how much from improvisation, we do not know.

We suspect that Homer and his fellow bards, devotees of or mouthpieces for Calliope (she of the fair voice) had at hand a wealth of set pieces in what must have been well trained and well stocked memories. In *The Origin of Consciousness in the Breakdown of the Bicameral Mind* Julian Jaynes argues that Homer and his fellows were not remembering and reciting as we would, but that they are actually being spoken to or spoken through by presences other than themselves, presences we would call "hallucinations" (Jaynes 1976: 371-72). When Homer is about to catalogue the ships and leaders of Agamemnon's army in the second book of the *Iliad*, he pauses for an invocation:

> Tell me now, Muses, dwelling on Olympos,
> as you are heavenly, and are everywhere,
> and everything is known to you—while we
> can only hear the tales and never know—
> who were the Danáän lords and officers?
> (Homer 1974: lines 567-71)

In effect, Homer says that we, as mortals, can only know by the indirection of rumors and stories; the Muses are immortal; they know by direct and infallible witness. The answer the Muses give, as Clara Park says, is "261 long lines crammed with 509 discrete facts which in a non-literate culture he [Homer] had no way of looking up" (Park 1991: 23).

There is controversy, of course, (and probably never to be settled) about whether Homer and friends actually heard voices or were spoken through as in hallucination; or whether they operated from memories similar to our fallible ones. In the latter case there is still controversy about how much of the *Iliad* was recited from memory and how much was woven together from formulas of improvisation; one estimate: "to quote Denys Page, 'about nine tenths of the Iliad's [sic] language was supplied by memory in ready-made verses or parts of verses'" (quoted in Park 1991: 24), and presumably Calliope's six singing sisters had their own mixes and balances between memory and improvisation.

The astronomy over which Urania presided was basically "philosophical"—addressing questions about the sphericity of the earth, debating whether the solar system was geo-centric or helio-centric, discussing and cataloguing the fixed stars and their positions, and recording the latter in at least one star catalogue of more than a thousand stars (1022, to be exact) that made its underground way from the famed library at Alexandria to surface in the library in Medieval Toledo. But for the most part Urania's scholars and students were involved in public discourse, not private publication in obscure academic journals. And as all voices and songs are substantial in our conscious experience of them, they are also fragile and evanescent when the empire strikes: so the Romans dismissed Greek astronomy as mere idle speculation in favor of astrology which, after all, had its practical applications such as enabling the adept to predict the future course of empire.

The Muses tend to be silent in the presence of causal materialism of the Roman sort. But the real provinces of the Muses in the Greek world were not the attic, the closet, the studio, but the agora, the theatre, the dance floor, places for casual or scheduled assembly and performance: the living word, sponsored by Mnemosyne, in the presence of living memories, and it is surprising what retentive memories for the spoken word the inhabitants of a basically oral culture will enjoy. During the Second World War, I found myself in command of an infantry platoon in basic training. That platoon was composed largely of non-literate draftees from the Appalachian coal fields. I don't say "illiterate" because that word has come to signify language-incompetence or language-deprivation, and those men enjoyed an extraordinary level of verbal competence. They never forgot anything the lieutenant said, from detailed orders to the most casual asides; everything said was recalled word for word, sometimes word for embarrassing word, but there at the ready, nevertheless.

And so to the *Museum* (The House of the Muses or Memory's House) in ancient Alexandria. It was apparently modeled on a much smaller Museum in Athens, and that one in turn appears to have been modeled on an even smaller one at Stagira, Aristotle's birthplace. The Museum at Alexandria was not just the fabled "library" as the Romans mistranslated the Greek word Museum (probably because the only things they could see and touch were the books). The Museum was a House of the Muses, a complex of buildings and spaces appropriate to the activities presided over by the Nine, including an observatory for Urania, a theatre for Thalia and Melpomene, a dance space for Terpsichore, public lecture rooms and commons rooms, and a portico, modeled on "the painted porch," a public ambulatory in Athens where Zeno and the Stoics taught in the first half of the third century BC.

The whole complex was a place of public assembly, much more like a university than what we would call a museum. Nowadays, we assume that museums will be secure, climate-controlled, well-lighted places where we store and exhibit things of art and the valuables and clap-trap of all kinds of history, but things in our museums tend to be kept in a sort of quarantine, cut off from the flow of public life lest they infect us with the give and take of discourse. The library at Alexandria was just one large space in the House of the Muses, but its fabulous reputation has swallowed up the rest of that House. It was established some time in the first half of the third century BC by Ptolemy Philadelphus or his father Ptolemy Soter and expanded to include between five and six hundred thousand volumes—with the proviso that the *Iliad* was counted as twenty-four books, as was the *Odyssey*—so, all told, there would have been between thirty and fifty thousand books as we would count.

About the library's sponsors and custodians, it is a pleasure to be able to report that Cleopatra VI whom we remember as a world class charmer was also a world class intellectual and bibliophile who played a key role in the development of the library's collection. That collection, if it had survived, would still constitute a world-class headache for the classics scholars and students in our midst. Instead, we survive to regret the depredations of the Romans who bashed the library about in their civil wars, and the depredations of the newly fledged Christians quarreling with book-burning intensity about their conflicting orthodoxies, and the final depredation, the Arab-Islamic conquest in 641. One apocryphal story (apparently designed to exonerate the quarrelsome Christians) credits the Arab leader who captured Alexandria with the final book burning: he is supposed to have torched the library on the argument that all the books in the library either agreed with the *Koran* or opposed it; if they opposed its truth, they should be destroyed; if they agreed, we already have the *Koran* so they're unnecessary.

In any case, as Clara Park has pointed out in an essay, "The Mother of the Muses: In Praise of Memory": "all classical literature . . . though it was transmitted in writing, existed in the sound of the human voice" (Park 1991: 31-32). Literature was "something to be listened to in public rather than scanned silently in private" (Moses Hadas quoted in Park 1991: 30). So that great collection of books was a collection of voices, stored in one corner of the museum's university spaces, and circulated by being read out loud. And I might add that the Arabic word *Qu'ran* (which we Anglicize as *Koran*) means a "recitation" as well as a "reading"; and as a further aside, I might add that reading remained a largely social activity, out loud and in company, until some time in the latter half of the eighteenth century. From an Elizabethan point of view there was something ambiguous, perhaps even unhealthy, about reading to oneself. When Hamlet is discovered "reading on a book" (Shakespeare 1992: II:ii:184), the Queen and Polonius take that behavior as a sign of his melancholy, and a bit later, Polonius applies that lesson. When he sets Ophelia to entrap Hamlet, he gives her a book to read, as he says, to "color / Your loneliness" (Shakespeare 1992: III:i:51-52), that is, to suggest that she, too, is melancholy, suffering from unrequited love and therefore desiring to be alone. A couple of lines later Polonius suggests another possibility, "Pious action" (Shakespeare 1992: III:i:54) because pious meditation was a healthy excuse for solitary, silent reading, And when Hamlet encounters her after talking to himself for forty lines ("To be or not to be," etc.), he assumes that she is praying: "Nymph, in thy orisons / Be all my sins remembered" (Shakespeare 1992: III:i:97-98). The hand-held book made its first appearance at the beginning of the sixteenth century, but even with the technology in place, it took over two hundred fifty years to evolve the flow of eye and the variegated humming in head and throat that we assume to be reading to oneself.

The Muses: they were the daughters of memory because without memory, no voice could be raised in song or story. And without memory, the listener could literally not hear the song or story. Homer tells the story of a bard, Thamyris the Thracian, who boasted that he could out-sing the Muses. The Muses did not kill him for that *hubris*, though as goddesses and daughters of a ruthless father they might well have done. "Instead, they 'in their anger struck him maimed, and the voice of wonder / they took away and made him a singer without memory'" (quoted in Park 1991: 26). In other words, they hit him at the source, at the well-spring of all utterance.

I pause at length over the Greek intuition that Memory is the Mother of the Muses because it says to me that memory is the Ur-art, not just the mother of the Muses but the spiritual presence that energizes their musings and their music. I don't mean memory mechanized as it was in the mind-factories of the eighteenth century, to the point where Blake felt called upon at the end of the century to dismiss the "Daughters of Memory" in favor of the "Daughters of Inspiration" in the "Preface" to *Milton* (Blake 1966: 411). I mean instead memory as that vast reservoir of what Santayana called "hushed reverberations"; and not just passive but active as in what Coleridge was to call the "esemplastic power," that which renders plastic, molds, unifies, gives coherence. Memory, in other words, *informs*, gives form to experience. As John Dewey said, "no experience of whatever sort is a unity unless it has aesthetic quality" (Dewey 1939: 40). Experience, Dewey repeatedly says, involves "a perceived relation between doing and undergoing" (Dewey 1939: 46, 50, 52), passive-receptive and active-informing. So memory as the Ur-art offers the constant possibility of an infusion of aesthetic potential; it is art-enabling.

Perhaps I should add at this point that I am thinking of art as conservative in the true sense of the word: that which ingathers, conserves, and in the process renews. I am assuming, in other words, that art is the new that renews, as against the present fashionable assumption that art should be at something called the "cutting edge," constantly rattling its cage from the inside with the bumper-sticker rhetoric that clutters the art-spaces of the contemporary scene, that clutter results in part from the mistaken belief "that it is the sacred duty of every artist to be radical in some way, whether it be political or philosophical or merely technical" (Graham-Dixon 1996: 126). As Robert Hughes puts it with his usual verve: "the American art world is swamped in a tide of political utterance that exceeds in sheer volubility and boredom anything produced by Popular Front Marxism during the 1930s or 1940s" (Hughes 1993: 43). And to that revulsion with art as politics we can add Lionel Trilling's caution: "Unless we insist that politics is imagination and mind, we will learn that imagination and mind are politics, and of a kind we will not like" ("The Function of the Little Magazine" quoted in Chace 1980: 17-18). Not art as fashionable clap-trap promoted by celebrities, but an art that brings us, in Whitman's phrase, "retrievements out of the night" (Whitman, "When Lilacs Last in the Dooryard Bloom'd" 1982), in that strange combined sense of something-found, something-made, an art that from Dewey's point of view offers "a remaking of the experience of the community in the direction of greater order and unity" (Dewey 1939: 81).

I don't mean to suggest that memory goes about dropping art works out of her sleeves but that, as process, memory is art-enabling. Henry James, in a letter of protest to H. G. Wells (10 July 1915), defined what to him was the function of art: "It is art that *makes* life, makes interest, makes importance . . . and I know of no substitute whatever for the force and beauty of its process" (James 1958: 267). And when I speak of memory as art-enabling, I am all but substituting memory for art in that definition: "It is memory that *makes* life, makes interest, makes importance," and that perception we owe to the Greek intuition that Mnemosyne is the mother of the Muses.

The House of the Muses at Alexandria aspired to encyclopedia, "instruction in the whole circle . . . or complete system of learning" (*Britannica* 1910-1911: IX.369). But since that museum's decline the provinces over which the Nine preside have, to say the least, shifted and changed: most radically, Urania's domain that has expanded to include the realm of the atoms and their particles as well as the realms of galaxies and black holes. And her language has changed. Her realms are so remote from the mid-world we inhabit that Euclid's language and geometry will no longer do; so Urania speaks in the elaborately evolved language of modern mathematics, a language which has changed much more radically than the language of poetic discourse which we can still manage to share, at least approximately and in translation, with Homer and Hesiod. Clio's province has been transformed into an empire, not just through the cumulative expansion of the kitchen-midden of written history but also through recent archaeological excursions into pre-history. And her empire has expanded to overlap Urania's by the inclusion of the historical sciences (natural history, geology, biology, and all their progeny). Clio must also take responsibility for our so-called "social sciences" with their shifting mix of history and story-telling, and their recent and increasing envy of the so-called "objectivity" of Urania's language and accomplishments.

If the novel is the heir-apparent of the epic, Calliopé's realm has also expanded, its borders blurred by overlaps with the provinces of Clio, Thalīa, Melpomene, Terpsichoré and Erato and all but dissolved in the realms of film and television. Melpomene's province has shrunk because we find it so hard to approach *catharsis*, that "clarification (or illumination) concerning experiences of the pitiable and fearful kind" (Nussbaum 1986: 391). In our world of causal materialism and in this century of massacre we tend to think of the evil, which might be pitiable and fearful, as something quantifiable—as, for example, we regard the Holocaust as transcendentally evil because it had such terrifying numbers. And the arguments about how that enormity should be kept alive in memory suggest that we need all the help Clio and Melpomene and their mother can give us, though perhaps a museum is not exactly the right answer.

The other Muses have had their provinces gerrymandered and for the most part underfunded. Recently Erato has been sent to the woodshed by angry congresspeople and other presidential hopefuls. Architecture and the visual arts continue to wander in search of a presiding muse, though they should be cheered by the fact that our word "mosaic" derives from a Greek word meaning "belonging to the Muses."

In our time, that is, since Shakespeare and Milton, the muses have evolved toward the singular. Shakespeare mentions the Nine only once (*Mid-summer Night's Dream*, 5.i.56); in "Lycidas" Milton calls on them as "Sisters of the sacred well" (15), and in *Paradise Lost* he transforms Urania into the "Heavenly Muse" who inspired Moses on Sinai (6-7) and to whom he prays for inspiration (Milton 1953). So our muse tends toward the singular as the *American Heritage Dictionary of the English Language* defines it: "the spirit or power regarded as inspiring and watching over poets, musicians, and artists"—as the provinces of Hesiod's Nine have expanded and overlapped into a transformed encyclopedia, so the Muses have expanded toward One. Or, as David Jones said in the "Preface" to *The Anathemata* (1952), "where there is order and sensitivity to the conserving of a form, there is the muse" (Jones 1952: 31, n3). "She," as he puts it, "that has music wherever she goes" (Jones 1952: 31), or, in Joyce's variant, "sea shell have music way river she flows."

The muse-singular reflects the several other dimensions of the word "muse" in modern English. As noun it means contemplation: "he is in a muse" (that is, meditating, abstracted from the passing scene); as a verb it can be either intransitive or transitive. Intransitive: to wonder or meditate, often upon something, as in Psalm 143:5, "I remember the days of old; I meditate on all thy works; I muse on the work of thy hands." Transitive as in Coleridge: "The maiden paused, musing what this might mean" ("The Destiny of Nations" 1796, 1817). Shakespeare uses the word in all three of its noun/verb modes, and that indeterminacy can, I think, carry us back to the confluence of doing and undergoing in John Dewey's definition of experience and by implication his sense of how memory informs experience. And from there I want to leap to a similar confluence of active/passive in poetry and the arts (and therefore in their mother, memory). Our words *poet* and *poem* come to us from the Greek verb *poiein*, to make; in Lowland Scots, poets are the *Makeris*, the makers. In Provence, the poet was troubadour from the Provencal *trobar*, from the Latin *turbare*, to search through, to search for, to find, and subsequently in

Provencal, to compose poetry. So poetry and the arts exist at a confluence of finding and making, where memory presides; where Goethe can say, "The songs made me, not I them."

Robert Graves attempts to focus on the mystery of that confluence in *The White Goddess*. He quotes part of a poem he had written about the Muse:

> *If strange things happen where she is*
> *So that men say that graves open*
> *And the dead walk, or that futurity*
> *Becomes a womb, and the unborn are shed—*
> *Such portents are not to be wondered at*
> *Being tourbillions in Time made*
> *By the strong pulling of her bladed mind*
> *Through that ever-reluctant element.*
> (Graves 1972: 343)

"The unborn are shed," that is, sent forth, given birth; and "tourbillions" are whirlwinds, vortices—little tornadoes in Time. Graves turns aside to explain: "time" he says,

> is not the stable moving-staircase that prose-men have pretended it to be, but an unaccountable wibble-wobble. . . . In the poetic act, time is suspended and details of future experience often become incorporated in the poem, as they do in dreams. This explains why the first Muse of the Greek triad was named Mnemosyné, "Memory": one can have memory of the future as well as of the past. Memory of the future is usually called instinct in animals, intuition in human beings. (Graves 1972: 343)

Graves then goes on to distinguish between dream and the poetic act: in the poetic act, someone is, or is supposed to be, in charge, but dreams are anarchic. He concludes that "[a] sense of the equivocal nature of time is constantly with poets" (Graves 1972: 344).

I think we can go on from there to suggest that there is an element of anarchy in memory, a shimmering sense of the equivocal nature of time similar to but less pronounced than that which we experience in dream. And some psychologists have recently speculated that perhaps the primary function of dream is the consolidation of memory, and so we come full circle to the observation that memory-time is nothing like an arrow.

At this point we might pause to ask just how close to our experience, how congenial to our thinking about memory is that Greek intuition that memory is the Ur-art; and at the same time we should pause to consider just how far, as memorists, we are from that non-literate Greek culture—or rather the literate but basically oral culture—that first developed and elaborated that intuition.

2. Comic Interlude

In one of the sustained hallucinations in James Joyce's *Ulysses*, Dublin, under the leadership of Leopold Bloom as "emperor-president and king-chairman" (Joyce 1986: 15:1471), is transformed into a utopia. In the celebration and general excitement of what follows:

> *The keeper of the Kildare street museum [that is the National Museum of Ireland] appears, dragging a lorry on which are the shaking statues of several naked goddesses, Venus Callipyge, Venus Pandemos, Venus Metempsychosis, and plaster figures, also naked, representing the new nine muses, Commerce, Operatic Music, Amor, Publicity, Manufacture, Liberty of Speech, Plural Voting, Gastronomy, Private Hygiene, Seaside Concert Entertainments, Painless Obstetrics and Astronomy for the People.* (15.1702-10)

You will notice that there has been an inflation from nine to twelve, but otherwise their hopeless urban middleclassness speaks for itself in the midst of the "placeless cosmopolis" which is the modern world.

3. The Australian Aborigines

I want to turn now to focus on the presences of memory among another and radically different non-literate people, the Aborigines of Australia. Much of what we know of them may be too little and too late. After the so-called "discovery" of Australia in the late eighteenth century, the colonists paid little attention to the Aborigines except to find them in the way and to treat them as an expendable nuisance. When anthropologists got around to them a century later, they found the tool kit and material culture of the Aborigines so under-developed that "impoverished" became the condescending label for that culture. Not until fairly recently have students of the Aboriginal culture come to realize just how rich in non-material goods that hunter-gatherer culture has been, and that realization has come somewhere on or beyond the brink of too late.

Linguists and historians can only guess how many languages there were among an Aboriginal population they estimate at about 300,000 in the late eighteenth century. The guesses range from 600 to 900 languages and dialects. This means, of course, that language groups must have been widely dispersed and, from our point of view, incredibly small. Yet the evidence suggests frequent contacts and interactions among language groups, and linguistic historians now assume that "multilingualism was the norm" (Sutton 6). But perhaps I should add here the suggestion that the sharp separation we assume from our experience of modern languages in our world may not have been anything like as sharp in that more fluid language environment?

We are used to the fact that what might be called our cultural encyclopedia (the total sphere of our knowledge of all things) requires vast networks of storehouses, libraries, archives, museums, databanks, and now (hurrah) the internet and cyberspace. And the maps that guide us among and through those networks are incredibly complex and flawed and easy to misread. Among the Australian Aborigines the cultural encyclopedia and its maps had to be held in the living memory of the language or hearth group, and for the members of the group to come into possession of that encyclopedia, or rather to be possessed by it, was "the highest goal in life" and was expected to take a lifetime, according to Theodore Strehlow (Strehlow 1971: 244). The outstanding student of what Bruce Chatwin called the Songlines of the Aborigines, their seemingly endless chant-poems, Strehlow added, all members of the group were expected to aspire to come into possession of that encyclopedic goal. The result was that individual language competence continued to grow and improve as members of the group aged, to the point where Aboriginal elders had acquired levels that among us would be, to quote Strehlow, "the privilege of the highly educated few" (Strehlow 1971: 247). The fragility of that encyclopedic possession means that we on the outside have to be particularly careful about the limitations of our knowing. Strehlow said in 1971 that for one important language group, the Aranda of Central Australia, "no one born after 1900 would have received full instruction" (Strehlow 1971: xxxv). That is to say, the passing of the cultural encyclopedia through living language, living memory from generation to generation had irreparably broken down by the end of the last century. And unfortunately it appears that all the other remaining language groups are similarly cast adrift toward decay and dissolution. Nevertheless, enough remains to us of this incredibly elaborated and articulated "Stone Age" civilization that we can sense the force of Joyce's pun: I am "astoneaged" (Joyce 1939: 18:15).

Thanks to their songlines, Aborigines could 'sing up' the countryside toward which or through which they were moving, even if they had never been there or seen it before. On one level the songlines were like intricate verbal maps of the countryside; on another level they were like cables woven around a mythic core, the presence in the landscape of evidence of the activities of the ancestral figures from what the Aborigines called the "Dreamtime," the time before time began, the time of Genesis. And time dissolves toward the equivocal here because the individual humming the songline can say: "as I am here now, I was there at the beginning because the beginning is here now in that outcrop of red rock where the dreaming ancestor turned over in his sleep." The multiple strands of those songlines, or rather those song-cables, had strands of practical information interwoven around the mythic and landscape core: migratory and food-gathering routes, where to find waterholes in what to our eyes would be a featureless and unforgiving landscape; when and where to apply the firestick to tame the land and induce new growth. The complexities of the Aboriginal firestick culture deceived the European settlers into thinking of the Aborigines as irresponsible wandering arsonists (see Stephen J. Pyne 1991).

Other strands in the songcables could include information about locations of and routes to ceremonial sites plus calendars and organizational plans. There were strands that told where in the geography of exogamy to seek a wife. Other strands could include the individual's ancestry and family ties, his inheritance, his identity, his passport (his personal songline) and where, in the rites of passage, he should make his Walkabout.

Collectively the songlines or songcables mapped the ways in which the spiritual identities of every individual and group were imprinted on and made manifest in the landscape. Those identities can not only be "sung," they can also be physically explored on the ground, and to explore them is to fulfill one's duty of, in Bruce Chatwin's phrase, "looking after the country" (Chatwin 1988: 14).

But in examining the presence of memory in another culture as foreign to us as that of the Australian Aborigines we create problems for them and for ourselves. Characteristically we seek to separate out the strands of their culture and categorize them: into songlines (their chant-poems), their mythology (the vast compendium of story that comprises their "history" and the "history" of all things), their dance and performing arts, body painting, painting on bark and rock, their sand paintings, sculptures, their ceremonies, rituals, cosmology, religion, social organization, and what it means to be a person. For the Aborigines themselves all those strands, which we analytically tease out and separate into discrete categories, are woven into one cable, the lifeline of articulate memory that sustains all things.

Strehlow repeatedly complains that the songlines cannot achieve truly great or "Homeric" status because he assumes that to be

Homeric poetry must exist "as an independent medium of artistic expression" (Strehlow 1971: 675). From his point of view the chant-poems are hampered by being immersed in what he calls "a composite art form." His complaint reflects our habitual impulse to analyze, distinguish, separate out into categories. None of the Aboriginal languages has, for example, a word for artist because when it was art-time everyone was expected to participate. Some individuals were undoubtedly more proficient at body-painting than others, but there were no specialist categories. What Strehlow misses in his pre-occupation with the poet as specialist is the extraordinary cultural achievement of the Aborigines. They have or rather have had until sometime in the recent past, the longest unbroken art tradition in the known history of the world, beginning at about the time the paint began to hit the wall at Lascaux or perhaps even earlier. And the songlines seem to match that extension in time.

In Joyce's *Ulysses*, Stephen Dedalus imagines the continuity of human flesh as a "strandentwining cable" (Joyce 1986: 3.37), braided of the navel cords of all humanity and offering a telephone line back to the Ur-telephone in the Garden of Eden. The telephone number is "Aleph, alpha: nought, nought, one" (Joyce 1986: 3.39-40), the initial letters of the Hebrew and Greek alphabets, signifying *in the beginning*; and "nought, nought, one" signifying Nothing and then Creation from Nothing as only God and (in Joyce's joco-serious imagination) the artist can create: the Earth, Eden, Adam and Eve. If we accept that strandentwining cable as extending back to the beginning of the Aboriginal (and human) heritage, then at its beginning and in its core we will find not the anthropological evidence of the tool kit strewn about in the Garden, but two interwoven strands, language and memory, strands which leave no bones to inform the archeologist but which tease us with the circular proposition: without memory language could not exist and without language memory, as we know it, would not exist.

Our analytic yardsticks don't provide very good ways of measuring the "survival function" of the various strands woven into those Aboriginal strandentwining cables, and we are liable to be put off by the insistent and overwhelming presence of nightmare in their mythology. But I suspect that the integrity of the whole cable is the key rather than the practical and social and personal and spiritual values of its individual strands. Figuratively speaking, the important thing for the Aboriginal groups creating and sustaining those strandentwining cables was their ongoing participation in the ropewalk of memory and experience that wove those cables generation after generation.

By the leverage of contrast we can see that we and our cultural encyclopedia are at the opposite end of a ropewalk. Where we are, the strands unravel into countless houses of memory. Faced with the impossibility of tracing all those strands back into their houses let alone weaving them into an integrated cable, many of us opt for a single strand, a specialization (poets here, physicists over there). Some of us, perhaps inadvertently, individually or in cooperation with friends and colleagues, manage to combine a few of the strands that continue to proliferate among us at such an alarming pace that we cannot hope to keep up. (And when I heard on the news the other night three people called "weed specialists" talking from three different universities about the present state of drought in New England, I didn't try to keep up, I gave up.) Programmatic attempts to invent an educational ropewalk that could provide a strandentwining cable for our time strike me as political and sterile, as the recent and frustrating arguments about core curricula in our colleges and universities demonstrate, or when we endure the folly of national discussions about what students in our high schools and colleges "need to know." The phrase "need to know" suggests content (a few pounds of physics, an ounce or two of poetry, and plenty of computer time) when what they really need are the verbal and mapmaking skills that form the basis and textures of human memory.

A parable to close with:

> In the 1950s Aborigines at Elcho Island off the north coast of Arnhem Land broke spectacularly with their tradition by building within the grounds of the Methodist mission a "Memorial" in which *rangga* (carved and painted wooden figures symbolic of world-creative powers) were stood upright for all to see. Legends inscribed on the surrounding fence imputed to named Aborigines credit for the Memorial and for benefits received from the missionaries. As *rangga* are customarily objects of revelatory rites to which men only are admitted, the Memorial was a place at which sacred things had been profaned by being brought out of the secrecy hitherto proper to them. (Maddock 1972: 1)

Apparently the Aborigines had reasoned: we have shown you our secrets; so now show us yours and level the playing field. One anthropologist suggested that the Western secrets which they wanted to have revealed to them were "education, economic security, employment and control over their own affairs" (R. M. Berndt, *An*

Adjustment Movement in Arnhem Land [1962] quoted in Maddock 1972: 1-2). For the purposes of my parable I take a somewhat darker view: that the secrets of our civilization have long since been published and hung out to dry. The Cross is above the altar; the Mezuzah is on the door post; the Crescent is on the dome of the mosque. The books are in the libraries; the equations are on the blackboard; the money is in the bank; the information is being processed on the internet. And today's secrets will be published on the internet tomorrow morning. The millions, perhaps billions of pages of classified documents in the bureaus of our security state apparatus are worth nothing in this kind of exchange. In sum, we can't respond to the extraordinarily generous gesture of the Aborigines because we've already turned our cultural encyclopedia inside out, and the secrets that remain to us are those that we don't know we don't know. They are the secrets that underpin Emily Dickinson's wonderful and terrifying aphorism: "Had we the least intimation of a Definition of Life, the Calmest of us would be Lunatics." And in her formulation, Definition, Life, Calmest and Lunatics are all capitalized for emphasis.

Chapter 3
Doing Memory and Doing Language

Toni Morrison, accepting the Nobel Prize for Literature in 1993 said, "we do language. That may be the measure of our lives." And for good measure I want to entwine that with "we do memory; that may be the measure of our lives" (Morrison 2006). Without memory there could be no language; and without language there could be no memory as we experience and know it. I don't mean to suggest that memory is basically verbal but that language enables the complexities and coherences that we know as human memory.

And language has been at the core of human evolution and memory since long before the beginnings of written history, since the emergence of so-called modern man thirty-odd thousand years ago. Privately I think that the emergence of modern man was not only a biogenetic event but also a language event, language born not gradually but in a thunderclap of evolutionary change. Linguists and linguistic historians tend to assume that language evolved gradually over thousands and thousands of years. Steven Pinker in *The Language Instinct* suggests "language could have gradually evolved" over a period of "350,000 generations" (Pinker 1994: 345-46). That fade-in is supposed to have begun with the language of practical application to and manipulation of the immediate environment. It is supposed to have begun with what I call "pass-the-ketchup," and then oh-so-gradually evolved toward the so-called "Universal Grammar" (Pinker 1994: 346) and symbol-juggling wonder world that modern man and his languages create and inhabit. I wouldn't dare print what I'm about to say or to say it too publicly because if I did I would be instantly excommunicant from the realms of academic intellectual enterprise. But I believe you cannot get to where we are from a pass-the-ketchup beginning. What pass-the-ketchup leads to is more passing and more ketchup. That, I suspect, is where the Neanderthals, if they had language, wound up, stalled in a culture marked by a tool kit that did not change over a period of 100,000 years. We can say "pass-the-ketchup" by grunting and pointing, but you can't say "we do language" that way, not in Toni Morrison's sense of the phrase.

It is my heretical belief that modern man began to "do language" (or language and memory began to do modern man) in an evolutionary thunderclap that took only a few generations, and in that language-beginning consciousness, as we experience and know it, had its beginnings. And to rub salt in the intellectual wound, I think that the inventors of language were children. For one thing because the grammar-generating center in the brain seems to close down at puberty and because recent researches into the development of pidgin languages and creole by the linguistic historian Derek Bickerton suggest that children from mixed language groups do the evolving of the pidgin and the creole in two or three generations. At any rate, once clapped by the thunder, language would have spread its side effects, among them the practical and manipulative skills that accomplished what one archaeologist, John E. Pfeiffer has identified as *The Creative Explosion* that took place plus or minus thirty thousand years ago, at the time of the emergence of "modern man."

Language in and of itself enables and confirms memory, and it should come as no surprise that well before the beginnings of written history human beings learned to discipline and re-enforce the language-memory bond through versification, recurrent rhythmic patterns and patterns of consonance (rhyme and/or alliteration) and the tonalities of chant or melody (see David C. Rubin). So we can remember that September, April, June, and November each have thirty days because that fact has been tucked into a rhyme and thus into folk-memory (sometime before 1590). Similar mnemonic aids have in the last several hundred years corralled the Ten Commandments, the Beatitudes, the Books of the Old and New Testaments, the early kings of Rome, the kings of England, the first twenty-one Presidents of the United States, and all sorts of other classroom chores and entertainments. William S. Walsh's *Handybook of Literary Curiosities* (1893) is a goldmine of this sort of "memoria technica." But we should not forget that such roadside rhymes can prove mischievous and at times maddeningly tyrannical, as with Mark Twain's nightmare experience of a rhyme that insisted on setting itself to the rhythms of the clickety-clack of a railroad train on a track: "Punch, Brother, punch / Punch with care / Punch in the presence of the passenjare." When he was traveling, he could not suppress or divert the rhyme, and its nagging presence almost drove him mad.

So our vernacular and popular culture is still aware of the mnemonic power of metrical speech and sing-song rhyme as were the preliterate peoples of the ancient world who trusted all kinds of information from the trivia of kitchen accounts to the high poetry of Homer to that sort of memory and retrieval. The Homeric poems were apparently alive and in performance in bardic tradition for several hundred years before they were written down. We know that Homer's heroes were Bronze Age berserkers, and that Homer viewed their actions from the age of iron. As an aside on bardic traditions: ancient Irish bards were trained to meet the expectation that they would be able to perform thirty-odd poems the length of the *Iliad* and the *Odyssey*, together with an assortment of shorter pieces. The Icelandic sagas existed in living memory for approximately 300 years before they were fixed in the amber of the written word.

We think that the Anglo-Saxon epic *Beowulf* was composed in the first half of the eighth century. The one dateable event in the poem, a Geat raid on the Franks, occurred in 520 when Beowulf, himself a Geat, was a young man. But lots of other historical, though undateable, events are alluded to in the poem, and the range of allusion suggests that the history of the sixth century was alive and well in the memories of the eighth-century audience. The manuscript of the *Beowulf* that survives (and barely) to us was written in the late tenth century, from what source or sources we do not know. But we do know that the manuscript drifted from a monastic library into private hands in the time of Henry VIII and that it was rescued for posterity through the good offices of a contemporary of Shakespeare's, Sir Robert Bruce Cotton (1571-1631), English antiquarian and founder of the Cottonian Library (now in the British Museum). The thread of manuscript survival has proven every bit as tenuous and vulnerable as we intuitively fear survival in bardic memory must have been.

But I want to leave the fractal history of manuscripts and books until later in this lecture and turn now to what Frances Yates, in a splendid and intensely scholarly book, has called *The Art of Memory*. In that phrase the term "art" drifts toward artifice, artificial memory—systematic procedures for ordering and retrieving vast stores of information and knowledge—and there is a conventional phrase that always follows that brief description: "without the help of the written or printed page." I want to suppress that phrase and point out that its negative preposition, "without," is ours. It did not belong to the original practitioners of the art of memory whether or not they could read or write.

The inventor of that art and the founder of the classical tradition (in legend if not in fact) was a Greek poet, Simonides of Keos (556-

468). As the story goes: he was at a banquet given by a nobleman of Thessaly named Scopas, and he

> chanted a lyric poem in honour of his host but including a passage in praise of Castor and Pollux. Scopas meanly told the poet that he would only pay him half the sum agreed upon for the panegyric and that he must obtain the balance from the twin gods to whom he had devoted half the poem. A little later, a message was brought in to Simonides that two young men were waiting outside who wished to see him. He rose from the banquet and went out but could find no one. During his absence the roof of the banqueting hall fell in, crushing Scopas and all his guests to death beneath the ruins; the corpses were so mangled that the relatives who came to take them away for burial were unable to identify them. But Simonides remembered the places at which they had been sitting at the table and was therefore able to indicate to the relatives which were their dead. (Yates 1966: 1-2)

Simonides noted that it was his memory of the *places* where the guests were sitting that made him able to recall their identities. Cicero continues the story:

> He inferred that persons desiring to train this faculty (of memory) must select places and form mental images of the things they wish to remember and store those images in the places, so that the order of the places will preserve the order of the things, and the images of the things will denote the things themselves. (Cicero, *De oratore*, II lxxxvi, 351-54 quoted in Yates 1966: 2).

So, the classical art of memory: specifically, memory trained to enable the orator to deliver long speeches "from memory with unfailing accuracy" (Yates 1966: 2).

If I wanted to practice that art (or rather that artifice) for this lecture, I would first imagine or recall a building with varied and interesting interior spaces. Inside the entry door on the left, I would position a photograph of Toni Morrison; beyond her, a free-standing sculpture of Janus with her face on both masks to image the language-memory twining, entwining. Beyond that a reproduction of one of the bison at Lascaux, an image of a man gagged, a poster-photograph of a bottle of ketchup. But this is beginning to sound like an Andy Warhol, so I'll stop, except to say that when it came time to give the lecture, I would stroll into the house and move past the images in my mind's eye and out would come the sequence of the sentences and paragraphs of the lecture.

Orators, lecturers, poets, and others trained their memories in this way in the Greco-Roman world, and thanks largely to Cicero that tradition survived into the Middle Ages and on into the Renaissance. En route the tradition was modified and embellished. Other spaces were added or advocated: a garden, a theatre, a tree or grove of trees, arrangements of ladders, the spheres of the planets, the houses and signs of the Zodiac, and so forth. And from late classical times the art of memory was intermingled with the mysteries of Neo-Pythagorean religion and that intermingling survived to surface in the Middle Ages in the Jewish and Christian cabalas and to blend into the Renaissance fascination with the occult—an interblending of the mysteries of the art of memory with the mysteries of the occult, a blend that was to send Giordano Bruno to the stake in the Campo dei Fiori in 1600. But that blend, further modified by an intermixture of religion, was to survive into the seventeenth century to color the origins of modern science in the persons of Leibnitz and Newton.

In the classical tradition the art of memory (artificial memory) was a discipline primarily cultivated for the practical purposes of public speaking (orations, lectures, poems, and eventually sermons), but the admixture of religion and the occult in the Middle Ages and Renaissance encouraged practitioners of the art to regard it almost as an end in itself. The art moved beyond immediate practical application toward the mysteries of encyclopedia, "instruction in the whole circle or complete system of learning" (*Britannica* 1910-1911). This reached a sort of climax in the Theatres of Memory in the High Renaissance, a climax of what Frances Yates has called the "Hermetic Cabalist tradition" (Yates 1966: 320). Guilio Camillo's Memory Theatre (Guilio Camillo Delminio, c1480-1544) was widely regarded as the climax of that tradition and its promise of ultimate knowledge. The theatre was conceived as an ampitheatre in which the memorist stood in the center of a semicircular stage, facing out toward a semicircular audience space of seven tiers, divided by seven aisles which radiated out from the Seven Pillars of Solomon's House of Wisdom positioned around the front of the stage. The forty-nine memory spaces thus delineated were further elaborated and subdivided into spaces marked by an intricate vocabulary of Hermetic signs and symbols, each endowed with allegorical and magical significance. On one level Camillo's theatre was to function as a system of memory spaces for use in the traditional practice of the art of memory, but it also encoded the magical and secret properties of the whole Hermetic Cabalist House of Wisdom. So it was at once a practical exercise in artificial memory and an encyclopedic vision of secret doctrine.[1]

---

[1] For a full analysis of Camillo's Memory Theatre, see Yates, Chapter VI. Editor's note.

And while Camillo and others were busy with their elaborate mnemonic secrets, the printed word (from the 1450s) was beginning its takeover of the western world. It is a commonplace (and, I think, a mistaken one) to say that printing took our memories away, but it is a commonplace that has been with us from the beginning of the world of print. The Renaissance scholar-editor Hieronimo Squarciafico thought he saw the handwriting on the wall in 1477, nine years after Gutenberg's death (1468?): "abundance of books," he said, "makes men less studious," and he said that just twenty-odd years before the founding of the Aldine Press in Venice and the advent of the relatively inexpensive handheld book. The books Squariafico thought too abundant were printed but they were still the outsized volumes chained to lecterns that we associate with monastic libraries. Squarciafico's anxiety seems to have taken two directions: one, that abundance of books would vulgarize scholarly enterprise and threaten the exclusive authority of the closed society of scholars. Literacy, heaven forbid, might spread and with it the rising tide that was to become the flood of Protestantism forty years later. And we should remind ourselves that it was the printing press, not laboriously handwritten posters that spread the wildfire of Luther's ninety-five theses. Squarciafico's second anxiety: abundance of books might well render men less diligent in their cultivation of the art of memory (if you had the book at hand, you wouldn't have to commit it to memory)—a misplaced anxiety as it turned out.

Erasmus, who was eleven years old when Squariafico voiced his worries, seems the embodied refutation of those worries. Erasmus spent some time as an editor in the abundance-of-books business in Venice, but in his scholarly mendicancy he traveled with a core collection of only thirty-two books. The extraordinary range of his scholarship, on the other hand, reflects a mnemonic grasp of thousands of books and manuscripts. There must have been a world-class memory traveling with that box of thirty-two books, but I should add that Erasmus, while he was a practitioner of artificial memory, was deeply suspicious of and averse to the sort of Hermetic elaboration being promoted by Camillo and many of his and Camillo's contemporaries.

But it is obvious that if printing hasn't taken our memories away, it has changed them, profoundly and in ways we have yet clearly to understand, let alone to master. Until a generation or two ago school children were expected and required to commit poems to memory on the vaguely articulated premise that the exercise would strengthen their memories and inform their minds with poetic valuables. Those memory exercises went out of fashion, rejected as rote-learning, rote-memory. Perhaps that was a mistaken shift in the winds of fashion, but the poetry I was required to memorize as a school child in Schenectady was, for the most part, second and third class stuff about poems not being as lovely as trees, etc., not informing the mind but cluttering it with cliché sounds and rhythms and sentiments. The in-and-out-of-fashion of those memory exercises is a measure of our uncertainty about how memory really works and therefore our uncertainty about how it should be trained—to the tune of a hickory stick or to the gentler sounds of the permissive classroom?

When we ask what print has done to our memories, we should hedge that question with a couple of qualifications: are we talking about our individual and personal processes of remembering, or are we talking about collective, social memory? If we are talking about individual and personal memory, we should be reminded that the intimate relation we assume between reading and writing and speaking was not necessarily the case in the Middle Ages and the Renaissance. Many who could read could not write and vice versa. Scholars could do both, of course, but they were few, tucked away in monasteries in the Middle Ages and traveling about in a sort of international private club in the Renaissance, and the language of that club was Latin, intensely so, to the point where a scholar like Erasmus had to turn down a lectureship at the University of Louvain in his homeland because he had no grasp of the vernacular language of his childhood; he could not, in other words, have lectured in what should have been his native tongue. Scholars apart, among the rank and file of literate people the two activities, reading and writing, were not always in tandem as we hope they are with us, and therefore reading and writing tended to stand in a sort of pre-literate relation to the exercise of memory. Memory was not, as it is among us, so closely involved with the written or printed word. So print took its time in colonizing the processes of our individual and personal memories.

As I've already said, the skill of reading to oneself in camera took centuries to develop into what we regard as the *natural* way to read, so natural that it can come as something of a surprise to be told by a research doctor at Yale Medical School: "And [reading] is the most complex of human functions. Reading is the pinnacle of what human beings can do" (Dr. Sally E. Shaywitz, quoted in *The New York Times*, [16 February 1995] A22). Surprise, because sharing that soliloquy with the printed page seems so natural, "so easy so to be" in Emily Dickinson's phrase.

> Your thoughts don't have words every day
> They come a single time
> Like signal esoteric sips
> Of the communion Wine
> Which while you taste so native seems
> So easy so to be
> You cannot comprehend its price
> Nor its infrequency
> (Dickinson 1970: 1452)

If we are to train our children's memories (and our own), it should be training in the skill of achieving and improving that "pinnacle of what human beings can do."

As for the collective, social memory, print and the consequent abundance of books has gradually and at the same time incredibly transformed the technologies of remembering. Historically, the continuities of collective, social memory have been subject to interruptions and severe displacements. The disintegration and scatter of the Library at Alexandria and the separation of the Eastern and Western Churches, effectively denied the heritage of classical Greece to Western Europe during the Middle Ages and the Early Renaissance, until that heritage began to filter in by way of the Arab (Moorish) culture in Spain and by way of the scholars who fled to the West with their books after the Fall of Constantinople to the Turks in 1453. In the early fourteenth-century Dante knew from secondary sources that Homer had existed, but he had no direct experience of the Homeric poems, and when copies of those poems arrived in the West, nobody could read them because Homer's language had been lost. It bore little relation to fourteenth century Greek and had to be reconstructed through a scholarly process that took literally centuries.

More discontinuities: the extraordinarily rich Moorish culture of Spain was symbolically centered in and built around the great Medieval libraries in Cordova and Toledo. Those libraries were broken up and scattered during Catholic Spain's violently intolerant re-conquest of the Iberian Peninsula. When the monasteries of England were dissolved by Henry VIII's royal decree in the 1530s, their libraries were scattered into not always bibliographically responsible private hands. And when Henry's son Edward VI established a Commission to reform Oxford University in the 1540s, the Commissioners undertook their mission with such zeal that they destroyed a great Medieval library in the process. Half a century later (in 1602) the library was reborn from its ashes under the aegis of Sir Thomas Bodley (1545-1612) after whom the library is now named.

And in all these major discontinuities and in a host of minor ones (a fire here, a night of pillage there) we have no idea (and certainly no catalogue) of what was lost in the shuffle. As for the missing catalogue, we do know that during the Middle Ages and the Early Renaissance there was no very firm convention about how to title a manuscript book of collected short pieces, let alone about how to catalogue it. So in Medieval monastic libraries books were in effect title-blind, a book labeled Bee-Keeping on its spine might well begin with such practical advice; but it might also contain, among other miscellaneous items, a manuscript copy of Aristotle's *Poetics*. And there was no indexing, let alone cross-indexing, except in the personal memory houses of individual monks and scholars. Print was rapidly to change all of that. Thorough and reliable indexes began to appear in books in the sixteenth century, less than one hundred years after Gutenberg. And cataloguing books in libraries by author, title, and sometimes subject, began soon after. But as recently as 1939 many of the college libraries at Oxford and Cambridge still catalogued their books in bound hand-written volumes by year of accession rather than by author or title; so if you wanted to have a look at Sir James Frazer's *Golden Bough* and didn't know the year of its accession you were in a fix. I found out about that fix when I was a student at King's College, Cambridge, in 1939. I discovered that it was easier to find the book by browsing the shelves than by rummaging through unalphebatized lists of authors and titles year by year.

Bound catalogues became more and more impossible because they went out of date so quickly as books proliferated toward overload. In the course of the nineteenth century the card catalogue emerged as the catalogue of choice to the point where the Library of Congress began to print and distribute book cards to all of its member libraries on request in 1901. But the drawers of cards are now being dispatched into the dustbin of history by the computer catalogue. So now we are comfortably (if a bit nervously) on line; and seated in my study at this keyboard I can log in to the Williams College Library a mile away. And I can also log in to most of the major libraries in the world. I can spend a few minutes in the catalogue of the Bibliotheque Nationale and drop by the Bodleian on my way home.

I pause here to wonder at this level of access, at how far it is from the closed libraries of the seventeenth and eighteenth centuries.

When Narcissus Marsh became Provost of Trinity College, Dublin in 1679 he undertook to reform the library and to improve access to its holdings, but he was frustrated by the Statutes of the College which dictated that only the Provost and the Fellows of the College could study in the library. Students were not admitted unless they were accompanied and supervised by the Provost or a Fellow. When Marsh became Archbishop of Dublin in 1694, he was determined to establish a free and open library in the Close of St. Patrick's Cathedral, and it is there to this day as Marsh's Library—though "free and open" is a quaint way to describe the library and its early practices because when a librarian handed a borrower the book he had requested, the librarian then led the borrower to a cage furnished with a table and chair and locked him in for the duration of his study.

But library catalogues are only one map-making technology that enables us to find our way in the maze of our social and collective memories. There are dictionaries that enable us to find our way around in all sorts of languages—from as early as ancient Nineveh when the Assyrians prepared clay tablets with lists of words pertaining to key functions of the government. Greek and Roman dictionaries tended to be less like what we would consider a dictionary and more like glossaries of unusual words and phrases, a thesaurus for the rhetorician. One outstanding Greek dictionary of the second century AD arranged the words according to meaning, a procedure I find intriguing and hard to imagine: how would I go about that in English? What categories, what hierarchies could I reasonably establish? And I am not ready with an answer to those questions.

Early dictionaries of English are strongly colored by the primacy of Latin as the language of literacy so that they are as much keys to Latin as to English. The first dictionary of English for English, the *English Expositor* dates from 1616 when modern English, thanks to Shakespeare and the King James Bible, was coming into its own. The great age of dictionary making (and, one assumes, dictionary using) began in the early eighteenth century with the appearance of a number of specialized dictionaries of the words used in different fields of study: geography, biography, history, philosophy, botany, zoology. And climactically there were Samuel Johnson's seven years of hard labor (with six amanuenses) which culminated in the publication of *A Dictionary of the English Language* in 1755. From Johnson's point of view the function of the Dictionary was not only to provide access to the words and their pronunciations and meanings but to stabilize the language; to, as he put it, preserve its "words and phrases from mutability." Johnson and his contemporaries were keenly aware of the history of the English Language and of how profoundly the language had changed in the 300 years between Chaucer and Dryden, and they were concerned lest similar change intervene to cut the future off from their own century. Alexander Pope shared that anxiety:

> Our sons their fathers' failing language see,
> And such as Chaucer is, shall Dryden be.
> Pope 1993: (1709) lines 482-83.

In short, Dryden (and Pope) would be all but unreadable, their language all but lost, by the year 2000 as Chaucer's language was to them. What Pope and Johnson did not foresee was that print itself, together with the broadening of the franchise of literacy, would have the effect of stabilizing the language, its spelling and pronunciation, if not always the meanings of its vocabularies of words and phrases. Encyclopedia originally meant, as we've seen, "instruction in the whole circle or complete system of learning" (*Britannica* 1910-1911), and it was to that whole circle that the Museum at Alexandria sought to address itself. Medieval encyclopedias were attempts at similar whole circles, full courses of instruction in the seven branches of the liberal arts, the *trivium* (grammar, logic, and rhetoric) and the *quadrivium* (geometry, astronomy, arithmetic, and music). But by the mid-seventeenth century encyclopedias were in transit toward the ordering so familiar to us, alphabetical arrangement by key words, names, or special topics. In effect, encyclopedias were being assimilated to the dictionary. Samuel Taylor Coleridge, when he outlined a project for an encyclopedia early in the nineteenth century, reacted against that alphabetical fragmentation of knowledge in favor of an exploration of and growth toward the whole circle of learning. But even in the early nineteenth century, Coleridge's plan overreached any possibility of achievement.

In the course of the nineteenth century two contrary impulses beset the makers of encyclopedias: one, a half gesture in Coleridge's direction, favored the concentration of related subjects in extended monographs; the other favored more and more refinement, subdivisions divided by subdivisions toward infinity. The first edition of the *Encyclopedia Britannica* (1768-1771) ran to three quarto volumes; the ninth, issued between 1875 and 1888 ran to twenty-four folio volumes, and an eleven-volume supplement began to appear in 1900. A German encyclopedia of science and the arts begun in 1818 set a dubious sort of record for thoroughness when it ran itself out at 167 volumes in 1890.

I should point out that, as maps of language, dictionaries are seriously flawed. They present language not as a guided and organic flow of grammar, syntax, and word, but as all chopped up, an alphabetical exercise in clerical isolation, frequently hobbled by a tendency to find synonyms and antonyms where they barely exist. Dictionaries enable our memories to encounter new words and their meanings together with the history of those meanings, but they do not enable the flow of language as in "we do language," which doing in turn finds its twin in the flow of memory.

Encyclopedias are similarly flawed as presentations of the whole circle of our knowledge, or as the Eleventh Edition of the *Encyclopedia Britannica* defines its "high ambition": to bring "all extant knowledge within the reach of every class of reader" (*Britannica* 1910-1911: I:ix). The alphabet has the effect not only of chopping knowledge up into bite-sized packages; it also tends to level the playing field by turning all the mountains into molehills and vice versa. And knowledge filtered through an encyclopedia's editorial bureaucracy usually lacks the defining or cutting edge of a point of view. Many of the articles in the 11th *Britannica* are the exceptions that prove this rule. They are still readable and useful precisely because they have articulate points of view. The articles were prepared by outstanding scholars who, in the community of scholars in that tight little island, knew or knew of each other. That community was still small enough to admit of cooperative acquaintance among editors and scholars. But the knowledge explosion since 1910 and the expansive fragmentation of the community of scholars have rendered the 11th's "high ambition" a dream of Cloud Cuckoo Land at our end of this century.

In addition to dictionaries and encyclopedias other ways of responding to what the abundance of books has wrought have evolved, other ways of making maps to extend memory's reach and our capacity to seek encounter with the whole circle of the collective learning of our time. There are concordances, alphabetical listings of the important words in books such as the Bible, listings that present the key words in the context of the phrases in which they occur. There were a number of concordances to various translations of the Bible prepared in the Middle Ages, but the great concordance explosion had to wait until the nineteenth century when the Bible was redone more thoroughly than ever and joined by concordances to Dante, Chaucer, Shakespeare, Milton, and even Tennyson in 1869 while he was still living and productive. And today we have the computer-generated word list which can offer the mind-numbing observation that the definite article "the" occurs 14,936 times in James Joyce's *Ulysses*, more than twice as many times as the indefinite article "a, an" which weighs in at 7051, a supreme example of the ways in which quantification can render all cats grey in the dark.

And since the middle of the sixteenth century when the mnemonic pressure of the abundance of books really began to be felt, there have been bibliographies: at first, attempts to list all extant and known works in Hebrew, Greek, and Latin (Conrad Gesner, *Bibliotheca Universalis*, 3 Volumes, 1545-1555). But the dream of a "Universal Bibliography" was soon overwhelmed by sheer numbers and abandoned in favor of national listings of books and listings under narrower and narrower subject headings. And by the late nineteenth century there were bibliographies of bibliographies.

What these exercises in quantities and in quantities of quantities tell us is not that abundance of books has taken our memories away but that abundance has challenged if not overwhelmed our capacities as makers of memory maps. I am reminded that one definition of a human being describes that being as the only animal that buys books it doesn't read.

But if we step back and attempt an overview of where we are in the mind-scape of our social and collective memories we might observe that memory does not stop at the water's edge. We are all familiar with the memory-support, the memory-extension that we expect and receive from family and friends. I don't remember but my wife, Honora will. And there are the elaborate extensions available through the twin formulae: "I don't remember, but I know who to ask" and "I don't remember, but I know where to look it up." Those who belong to the print-has-taken-our-memories-away party regard that latter formula with distaste as though those of us who fall back on knowing where to look it up are suffering from some sort of moral weakness. The aristocratic genius-hero of Ford Madox Ford's tetrology *Parade's End* looks down upon and despises the middleclass habit of looking things up in an encyclopedia, and then he is humiliated when the stress of the trenches of World War I wipes his memory-slate clean, and he is reduced to reading the encyclopedia to restock his mind with the general knowledge that he had once regarded as his birthright. But I'm not inclined to feel guilty when I look something up because I've spent much of my mature life remembering or trying to remember where to look, or stumbling by accident or serendipity into a place where it might be found or looked up. The 9000-odd three-by-five cards that went into the making of my annotation of Joyce's *Ulysses* are for me a career landmark in the mnemonic enterprise of I don't remember but I know or will try to find out where to look it up.[2]

The apparent dead-end of encyclopedic overload which seems to threaten us turns me once more away from imagining memory in terms of finite storage and retrieval and back toward the Greek intuition that memory is the Ur-art, the mother of all the arts including the art of map-making. And that art can still enable us to string the personal and individual clues or threads that make it possible for us to find our way around in the labyrinthine encyclopedias (circles within circles within circles) which are the social and collective memories of our time.

[2] Gifford 1988: Editor's note.

Chapter 4
The Intertwining of Language and Memory

Back to the twining, the intertwining of language and memory: as we do language, so we do memory; as the newborn child comes into language, so it comes into memory and vice versa. But what is the primary function of language that the newborn should be so well-prepared for entry into it? One of the great mysteries (not to say one of the great miracles) is the complexity and speed of that entry.

For centuries it has been assumed that the primary function of language resides in its practical and manipulative applications, in the ways in which it enables human beings to coordinate with and order their natural and human environments; that is, the primary function of language has been assumed to be what I call pass-the-ketchup. As I said in the third lecture, I don't think language could have evolved from pass-the-ketchup to language as "we do it." In the last century and a half pass-the-ketchup has added a dimension called "communications"; and communications has been developed to the point where there are communications consultants all over the place, and communication is enshrined in departments of communications in many of our colleges and universities. Language-as-communications is urged on us by the near fabulous developments in communications technology, from the telegraph in 1844 to the internet and cyberspace today. That technological revolution has reinforced the mechanical expectation that messages can be sent and received as sent. Fax it to me.

The instant and exact replication of which our machines are capable encourages us to expect an analogous capacity for sure replication in language itself between and among human beings. But the analogy is false and dangerous. No such replication (message-received-as-sent) is possible in the to-and-fro of everyday language, let alone in political discourse, let alone in poetry. If exact replication is what language is (or should be) for, then language as we experience and know it is a most uncertain and fallible instrument, deserving of all the maltreatment it gets from the deconstructionists who don't believe that anything can be said and understood as said at one end of the spectrum to the military jargoneers who want us to call a battlefield a “violence processing area” at the other end. As we all know, that’s the end where “collateral damage” means that some civilian bystanders have been blown away.

I suggest we turn the tables on pass-the-ketchup and communications; that we say instead: message transmission and managing the necessities are important *but secondary* functions of language. The primary function of language (its real survival value) is to make memory (and thus anticipation) possible, to make what we know as consciousness possible, and with it, continuity, community, communion. Not that memory and consciousness are verbal—far from it—but that memory and consciousness as we unselfconsciously enter and experience them are not possible or thinkable without language as the enabling medium. As a teacher of literature I had to remind my students over and again that while the medium of poetry and of our discussion of it is words, the experience of poetry in its full impact is non-verbal; something more than “just words.”

If we monitor the language that swirls around us and flows by us each day, we discover that little of it is “useful” in the pass-the-ketchup, communications sense. I once had a friend who dropped out of graduate school to take a job as technician (filming, tape-recording, etc.) with an expedition that was studying the !Kling Bushmen of the Kalahari Desert. I wrote to ask him what their primary use of language was. In reply, he estimated that 96% of their use of language was for narrative, an endless storying of their lives. And they never repeated the same story twice; to repeat a story was to make an outrageous joke, and the play-back of a tape-recorded story would send them off into seizures of laughter.

Even in our “high civilization” most of our everyday use of language is to absorb, to touch, to animate, to shape. We all, individually and collectively, story our lives, our histories, our worlds. When that is said, one skeptical reaction is to observe that in storying our lives we excuse and screen ourselves, cut ourselves off from reality. But the opposite is always possible: if the storying is well-crafted, well-done, it should put us more firmly in touch; it should sharpen our sense of the rhythms and contours of our lives. In effect, I story or sing or stage my experience in the echo-chamber of the self in order to re-enjoy, re-suffer, re-puzzle, to clarify, to shape, as well as to share, to amuse, to teach. Deceptions are always possible, but they are the risks we must suffer as “we do language” if language is to enable memory, consciousness, community, communion—because the primary route language follows is not the direct route of pass-the-ketchup and communications, but the indirect route toward what Oliver Sacks has called "the artful scenery and melody of experience and action" (Sacks 1985: 141).

The way children enter language re-enforces this intuition that the primary function of language is the poetry or melody of memory and consciousness. Not long ago we were at dinner in a restaurant in New Orleans. Among us was an eleven-month old child in a sort of plastic recliner on the table top. She was gazing up at an immobile ceiling fan, and then she pursed her lips as if to blow. She was symbol-making, linking the fan's potential for breath with the gesture of her mother blowing gently on her face to amuse her. She was showing what the psychologist Jerome Bruner has called "a set of prelinguistic 'readinesses for meaning'" (Bruner 1990: 72) and the meaning she was ready for was not turn-on-the-fan but metaphor: the fan's breath is like my mother's breath on my face.

We commonly speak of children as learning or acquiring language. But Noam Chomsky and his associates and followers have been arguing toward the position that the human brain is hard-wired for language, that there is what one of those followers, Steven Pinker, has called *The Language Instinct* in a book with that title. Yes. But I prefer a less hard-science way of putting it: that the newborn child is preconfigured to language and meaning and thence to memory. The psycholinguist Roger Brown likens the process of a child's coming into language to "the biological development of an embryo" in the womb, and other researchers independently use the same metaphor, according to Ursula Bellugi (Sacks 1989: 84).[3] Recent studies suggest that late in the third trimester the unborn child is aware of the tones and cadences of its mother's speech and that it listens differently to speech than it does to other sounds such as its mother's heart beat (Anthony DeCasper). Roger Brown observes that at eight months children can "vary their intonations in mimicry of adult speech" (62), and when speech begins at around eighteen months, it isn't as though the child were learning the rules by rote or by imitation but by half-

[3] In *Psycholinguistics: Selected Papers*, Brown argues that "The very intricate simultaneous differentiation and integration that constitutes the evolution of the noun phrase is more reminiscent of the biological development of an embryo than it is of the acquisition of a conditional reflex" (97). Gifford quoted from an essay in the *Harvard Magazine* which I have been unable to obtain. This chapter in Brown's book adapts a paper from a special issue of the *Harvard Educational Review* 34.2 (1964), 133-51 and appears to include much if not all of the material from the *Harvard Magazine* essay. Editor's note.

discovering them and half-making them up as it goes along. (And it is well to remind ourselves that grammar is not a fixed and inflexible system of rules, as by analogue to the fixities of Latin grammar, we once thought it should be, but that grammar is a history, a living historical stream.)

Perhaps it would help our understanding of a child's coming into language if we changed our terms. Instead of saying "learns"—with its implication of "by rote"—instead of saying "acquires"—with its implication of something grasped and stored away—we might try "infused with," "entry into," "immersion in." "Come on in, the water's fine," remembering as we say it that day-old babies can swim, that they are born knowing the "kinetic melody" (in A. R. Luria's phrase) of swimming but will forget the melody and have to relearn later in life. Just so, the infant is already infused with the melody of speech and is ready for immersion in language.

Linguists once thought that the child began by "learning the rules" in order to make sense; and later learned to break the rules in order to make metaphor and nonsense. But as they enter the amniotic fluid of language, toddlers behave as though conforming to and breaking the rules were two aspects of the one process of discovery-invention. Sense-making and nonsense-making, or, more precisely, literal sense-making and metaphoric sense-making are generically one and the same. As Oliver Sacks puts it, the little child has "a genius for language, between the ages of twenty-one and thirty-six months . . . which ends at childhood's end," at twelve or thirteen, and that is "the only period when the brain, from scratch, can actualize a complete grammar" (Sacks 1989: 83).

If that window of opportunity is missed, the consequences can be, to say the very least, severe. Oliver Sacks reports the case of an eleven year old deaf child who had never entered language; the child had no "clear sense of the past"; his awareness lacked "autobiographical and historical dimension" (Sacks 1989: 40). Sacks also cites A. R. Luria and Ia. Yudovich's account of language-retarded twins: "their play was repetitive and uncreative." When they were released into a normal language system,

> the whole structure of the mental life of both twins was simultaneously and sharply changed . . . and after only three months we observed the beginnings of meaningful play . . . the possibility of productive, constructive activity in the light of formulated aims . . . intellectual operations which shortly before this were only in an embryonic state . . . . (Ellipsis in the original; quoted in Sacks 1989: 42n)

There are relatively few recorded cases of total languagelessness such as the Wild Boy of Averyon, but the clear intuition, based on sparse and sketchy evidence, is that languagelessness sustained beyond puberty would be an irreparable, dehumanizing disaster for the child. The recent account published in *The New Yorker* of a California child who had been condemned to an all but languageless solitary confinement makes it clear that the child has been severely impaired, and the prognosis for some degree of recovery is not very promising. The experiment that would demonstrate the disastrous consequences of language-less-ness is, of course, the Forbidden Experiment, but there seems little doubt about what the outcome of such an experiment would be.

Language is not just a word-heap, though the 600,000 word horde of the new *Oxford English Dictionary* might tempt us to think so. The fundamental basis of language is not the word but the grammatical flow of sentence and proposition, the word enfolded. And as Susanne Langer said of language, "metaphor is the law of its life." She continues, "[Metaphor] is the force that makes it [language] essentially *relational*, intellectual, forever showing up new, abstractable *forms* in reality, forever laying down a deposit of old, abstracted concepts in an increasing treasure of general words" (Langer 1951: 141). Proposition, sentence, and metaphor return us to the intuition that the primary function of language is the poetry of consciousness rather than communications or pass-the-ketchup.

Oliver Sacks remarks: "It seems clear that grammatical potential is present, even explosively present, in every child's brain."[4] That word "explosively" reminds me of Helen Keller's explosion into language. As she approached the age of seven, she was locked in a sightless, soundless, languageless void; her destructive impulses and tantrums became increasingly dangerous to her and to those around her. Persistent attempts to teach her language in pass-the-ketchup school ended in failure, frustration, and tantrum. The breakthrough into what she was later to call "the mystery of language" came when her new teacher, Annie Sullivan, held one of Helen Keller's hands under water flowing from a pump and sign-spelled W-A-T-E-R in the other palm. Helen Keller says "I knew then that w-a-t-e-r meant the wonderful cool something that was flowing over my hand" (Keller 1974: 34). Language

[4] Probably quoted from *Seeing Voices* but I have been unable to locate the exact place. The closest I've come is "[t]the child is not taught grammar; nor does he learn it; he *constructs* it . . . . And this would not be possible were the grammar, or its possibility, not already within him, in some latent form that is waiting to be actualized" (*Seeing Voices* 82). Editor's note.

and what she called the awakening of "my soul" were entered through the aesthetic excitement of tactility rather than through the hand-to-mouth utility of a request for a glass of water.

Annie Sullivan's letter of witness to that event (5 April 1887) describes the aftermath of excited discovery and peace of that day, her charge transformed from demon into "a radiant fairy" (Keller 1974: 274). And a radical furthering of that explosive release into language came when Annie Sullivan realized that her seven year old student was in language terms just approaching the age of two, so she abandoned formal classroom sessions in favor of an induction into language like the one being used with Helen Keller's fifteen month old cousin. Come on in, the water's fine. Within two weeks Annie Sullivan was reporting extraordinary success: Helen, she wrote, "learns because she can't help it, just as the bird learns to fly" (Keller 1974: 275).

Hilde Schlesinger, in a study of mother-child behavior, observed that "some [mothers] talk *with* their youngsters . . . some primarily talk *at* their children" (quoted in Sacks 1989: 65). Some mothers, in other words, foster growth; others seek control. Annie Sullivan had the intuitive wit to switch from trying to control to fostering growth, and in so doing she enabled a remarkable woman to complete the embryonic growth of what she called her soul and to bear witness to the uniqueness of the process of coming into language, into meaning, into memory.

My argument is not that the process is sequential: first language, then meaning, then memory, but circular; all at once. Psychologists exploring the development of memory in little children distinguish at least three different kinds or levels of memory. "One is 'generic' memory in which the most general attributes of a familiar situation are stored," such as the memory of a familiar place, a childhood home or homes. "When an event has occurred in a child's life about five or more times, it tends to be stored in this general [generic] form" (*The New York Times* [6 April 1993]: C11). "A second kind of memory, 'episodic,' is for specific events," memories of "a distinct event at a given time and place," something unusual and special; and as the child begins to acquire the grammar and art of narrative, episodic memories begin to be woven into "autobiographical memory," the third kind, usually at about three and a half years old (*The New York Times* [6 April 1993]: C11).

Little children are possessed by a "'readiness' for narrative" (Bruner 1990: 45) just as they are possessed by a readiness for meaning and for language, but they also have to learn the skills of narrative through that endless question and answer chat in which adults review events with little children and shape events toward anecdote, toward narrative. What did we do yesterday? Circus. What did we see at the circus? Elephant. What did the elephant do? . . . Did you laugh? Did you cry? So the child learns to evaluate mini-events, to choose key ones and shape them into a story with a beginning, middle, and an end. And that art, that skill, blends toward autobiographical memory; and, psychologists say, the events we as adults remember from early childhood are those we have most thoroughly and firmly storied into narrative for others and/or for ourselves. "As autobiographical memory begins," they say, "ends" (*The New York Times* [6 April 1993]: C11).

But I should interrupt here to point out that it is precisely in this realm of encouraging and developing the skills and fun of narrative that television is a lousy baby-sitter because TV is not a narrative medium; it is a visual medium dominated by gesture. It cannot question and answer, and thus shape the responsive flow of story.

The progression which psychologists studying the development of memory in early childhood have outlined moves from generic memory to episodic memory and thence toward narrative and autobiography. And, the argument goes, children's early episodic and narrative memories fade over time and survive only rarely through autobiography into adulthood. Thus adult memories of early childhood do not provide reliable data for the study of the processes of memory in childhood itself. The curious thing about the progression from generic memory to narrative and autobiographical memory as outlined is that adult recollections of early childhood are so dominated by a sense of place. I have at hand at home a bibliography of 200-odd autobiographies in almost all of which a sense of place is the memory-rich lodestone of early childhood. I know that to be true in my case. A few years ago I had occasion to lecture on topos, a sense of place, for a class in environmental studies. I knew the place-rich emphasis of those autobiographies, and thought to check those accounts of memories of place in early childhood on the pulse of my own memory.

I began to explore the presences in my memory of the two places where I spent my early childhood: winters in the upstairs flat of a two-family house in a working class neighborhood of Schenectady, New York; summers in a cottage (called a camp) on an acre of land on the eastern shore of Ballston Lake, eleven miles to the north. Summer memories seemed the more place-rich; winter memories the more people-rich, so I focused on summer. I started with a United States Geological Survey map of the lake (Round Lake Quadrangle), and then I realized that I was beginning at the wrong end—with the region that was available to my canoe and bicycle when I was eleven or twelve. So I began again, sketching large-scale maps of the acre and its surround, floor plans of the cottage and its outbuildings, plus rough sketches of aspects of the landscape, not a particularly photogenic landscape but a place nevertheless. The floodgates opened, to the point where I felt threatened by a mnemonic affliction, the urgencies of early memories in advancing age, and I had to say: "Hey wait. I'm not writing reminiscences of early childhood; I don't want to get lost in what James Merrill has called 'childhood's inexhaustable brain forest.' I'm trying to address the question of a sense of place: what is it like to be emplaced?"

In the process I discovered that (at least for me) a sense of place was extraordinarily close to the "roots of memory." So what is "generic" and in the background for psychologists examining the growth of young children's memories is central and seminal to at least this adult and (I suspect) to a host of others. I dwell on this because I wonder whether the grounding provided by a sense of place isn't basic and general in coordination with the body as the mind's ultimate frame of reference (Rosenfield), perhaps more important than these studies with their focus on narrative and the mnemonics of time have allowed. Granted that narrative is the gateway to what Jerome Bruner calls "Acts of Meaning," I still hang out around a sense of place.

This whole question of the relation between place-memory and time-memory may be an important open question because the suburban cosmopolis that we inhabit is so essentially placeless. We might well be concerned that that placelessness is depriving the children among us of a ground bass that may be essential to the growth of healthy narrative and autobiographical memory.

Just as language has grammar, narrative has grammar, an interweave of structural and relational components that underpin and energize its flow and its coherence. There is, primarily, sequence and the question of what determines sequence. We think immediately of chronological order as the conventional determinant of sequence. But when you listen to the improvised stories of the day to day, you begin to realize that interruptions and asides, intrusions of collateral information, backtracking into parallels and contrasts are closer to the rule. Scrupulous chronological order emerges as the highly skilled, one might say "professional" exception. In addition to sequence, we assume that narrative will be concerned with an agent or agents, some sort of agency that usually has a purpose, a goal. And a third aspect of the grammar of narrative: there is the focus provided by an identifiable point of view; someone is doing the telling; someone is the voice that makes it hang together, that makes it make sense.

As narrative memory absorbs episodes, the child begins to recognize that different people see things differently; the child begins to acquire "the cognitive ability to see an event from several points of view" (Bruner 1991: 33). This strengthens and deepens autobiographical memory, and the child embarks on the life-long process of composing "the rough and perpetually changing draft of our autobiography that we carry in our minds" (Bruner 1991: 33).

As the child grows older, the narrative scheme thickens. The child learns to distinguish between everyday, conventional, normative behavior and exceptional or eccentric behavior. "When people go into the post office, they behave 'post office'" (Bruner 1991: 48); and if they don't, if someone behaves in an exceptional or difficult way, the child not only takes note, but also feels the need for narrative, the need to story that notable behavior, to assimilate or accommodate it into the expanding web of social narrative which is the child's emerging world view.

In the process of coming into narrative, the child is "enormously aided" by the family's (and gradually the neighborhood's and the village's) "stored narrative resource" (Bruner 1991: 67-68) if there is any village in this village-less coast-to-coast suburb. But those resources will, as memory does, always retell their stories tailored to the present tense. The story of my parent's courtship remained the same through each retelling right down to my mother's line for the final curtain (which, of course, proved temporary): "I could never marry a freak like that." But in each retelling, the story was slanted toward the present tense of its relevance to what was going on in the here and now.

I cannot, for example, recall my experience of first grade without the coloring and focus offered by the lens of retrospect. At some point late in her life when I was away at college or at the wars (I can't remember which), the teacher of that first grade class confided a ring to my mother. My mother was to pass the ring on to my future wife. That (and the subsequent career of the ring) modifies the whole view my mind's eye has of that child and that teacher in that classrooom. As one psychologist has remarked, "We are in the middle of our stories and cannot be sure how they will end; we are constantly having to revise the plot as new events are added to our lives" (Donald Polkington, *Narrative Knowing and the Human Sciences* quoted in Bruner, 1991: 116).

As in memory, so in narrative, "truth and possibility are inextricable" (Bruner 1991: 53). Our memories are anything but comprehensive and exact, so we fill in the interstices of inexactitude with the flexible mortar of narrative, perpetually creating what one psychologist has called "'sympathetic weather' in our memory reconstructions" (F. C, Bartlett quoted in Bruner 1991: 59). I'm not talking about falsehood or falsification but about memory and the narratives in which it is embedded. For example, when John Dean, President Nixon's aide, was recalling the Nixon White House during the crisis of the Watergate Hearings, he came across as someone with a tape recorder's capacity for the memory of conversations. But when the real White House tapes were revealed and compared with transcripts of Dean's testimony, most of his recall of who said what and to whom was far from accurate though throughout he had gotten the gist of what went on right. His memory was accurate in general, much less accurate in the particulars.

As the psychologist and educator Jerome Bruner has remarked, personal narrative has a "shadowy epistemology" (Bruner 1990: 55). What we can know of our past is alive and, in ever-shifting ways, well, but it is not as dependable as judge and jury might wish. I once put the question to my journal: what was the story of my adolescence? There were lots of fragments, snapshots and bits of talk, fractured threads of narrative, but the timetable was so moth-eaten that it could never have supported a story line. That line would have to be invented now and imposed on the unruly scatter of recall in retrospect. I could get a general feel for the drift of this or that period, and plenty of seeing now what I had not seen then and the consequent shards of guilt (the gift that lasts a lifetime). And drift is the word: when there were choices to be made, they were made by drifting or slipping through rather than by definitive choice. But if I were to organize the story now, the drifting would sound a good deal clearer and more decisive than it ever was on the ground.

And so back to the narrative resources which the village (if there is one) and the community have to offer the growing child through their wise (or unwise) people, through their schools and their cultural and religious institutions as the child begins to come into "that central synthesis of act and thought and value, of history and poetry and philosophy that . . . [the Greeks] called . . . *paideia*" (Park 1991: 41). As Hugh Kenner describes it: "that elusive incorporation into one's mind of the viable past, to draw near which is to incorporate that which we are" (Kenner 1968: 130). In other words, as memory grows, the young begin to enter into the "sign-stream" of our culture. As the anthropologist Clifford Geertz has said, "We are, in sum, incomplete or unfinished animals who complete or finish ourselves through culture" (quoted in Sacks 1989: 10). And in these so-called multi-cultural times, we have to remind ourselves that just as we have no choice but to be born, we have no choice but to be *born into*, a family, a language, a memory, a neighborhood, a culture, a sign-stream. We can, of course, test the waters of other languages, other cultures, other sign-streams, but we are not born into a supermarket where there is a choice among off-the-shelf cultures, or a choice among (as the unfortunate phrase goes) off-the-shelf "life styles."

Another way of approaching this point is to observe that "civilization is memory" and that our personal memories are the possibility of the ur-art within us, an art that makes us—at least potentially—active agents capable of participating in our civilization. A lot has been made in the last century (with a lot of moaning and groaning) about the disruption or loss of a settled culture and ideology, a lot of lament about our not being born into a stable and coherent civilization. T. S. Eliot thought that the destabilization had happened at around the time of Milton. Allan Bloom in *The Closing of the American Mind* thought it had happened at the beginning of this century, and that it had reached a climax in what he regarded as the supremely disorienting decade of the 1960s. What T. S. Eliot had overlooked was that while Dante and Milton had been born into what look to us like stable and secure cultures, they both were mavericks, ideological and political outlaws. The Florentines were all set to burn the exiled Dante at the stake if they could only lay hands on him, and Milton, after the Restoration was imprisoned and in danger of his life, though the authorities relented and sent him, blind and impoverished into what amounted to internal exile. As for Allan Bloom, there's no American mind to be seen behind that closing door. It's all Nietzsche and Plato and chaps like that; no Jonathan Edwards, no Ralph Waldo Emerson, no Nathaniel Hawthorne, Herman Melville, Walt Whitman or Emily Dickinson, only a couple of snide misrepresentations of Henry David Thoreau to demonstrate that Bloom had never read him.

I suspect that the farther shores of past civilizations look invitingly stable to us because that's the way we have composed them, the way we go on composing them. We think of Thoreau in the quietude of his cabin on Walden Pond because that's the way he composed it for us in the book called *Walden*, published seven years after he had plunged back into what he had never left: the snow-balling impact of the industrial revolution and the storm and stress of the anti-slavery movement and the run-up to the Civil War. We think of Wordsworth as secure in the quietude of the Lake District, not knocked about by the French Revolution and its aftermath in Europe and Britain as he was. In effect, we envy them what we have composed as the realms of stability in which they lived, and that envy reveals our desire for a world that would hold still long enough for us to be able to decide whether to stay on or get off.

Paideia: "that elusive incorporation into one's mind"—the operative word in that definition is "elusive" because that which is to be incorporated, "the viable past," refuses to hold still; and meanwhile we are caught in the flux of choice because "viable" clearly requires us to evaluate and to be active agents in that incorporation, not just the passive recipients. I see that wish that things would hold still everywhere about us, for example in the national attitudes toward and expectations of our educational institutions: they should get on with it and turn out by age 22 (if not by age 18) young adults capable of gainful employment. If we ask, what about *paideia*, that is shrugged off as a waste of time in a vocational world, and anyway it'll happen willy-nilly and of its own accord because we're all Americans. Underneath that assumption is the assumption that there is such a thing as a fixed maturity, an age when that elusive process of incorporation must have been (and of its own accord) accomplished. Our whole society is geared to assumptions like that: the age of consent, drinking age, voting age, draft age, commencement and the diploma in hand, and beyond those milestones, a stable maturity, a level road that should lead happily into the ever-after. Unfortunately, the road isn't like that. The journey toward *paideia* is never over and done with; so much is that journey still in progress that I often wonder (and with some anxiety), "Isn't it about time to begin?"

I see evidence of this desire for a fixed and stable world in the recent popularity of the concept, Post-Modern. My desk dictionary defines modern as: "Of or pertaining to recent times or the present, not ancient," and it derives it "from Late Latin *modernus*, from *modo*, "just now," originally "exactly," "to the measure," from *modus*, "to the measure." The prefix "post" "indicates 'after in time; subsequent to,'" so Post-Modern means after recent times, after the present, or in the future–near or far is not clear—but some other time when history will be holding still. A student of mine, when she first heard the term fifteen or twenty years ago, quipped: "Post-Modern; so then we're all dead." But seriously, does the term mean after my time? after our time? after what Francis Fukuyama (conservatively and perhaps prematurely) had pronounced "the end of history," when the Soviet Union came unstuck and we had won the Great Potlatch?

But even more seriously, the term Modern itself is adrift: if it means in my time, in these times, where are the bounds and limits? Does "in my time" mean in this decade? in my lifetime? in the twentieth century? Does it mean in the period since the Civil War as in "modern

America" or in the period since 1776? since the Enlightenment? since the Renaissance? "Modern English," for example, means the English Language since the early sixteenth century. Or does modern begin with the Fall of Rome? or the beginning of written history? or the time since the emergence of "Modern Man" and the birth of language, memory, and art in "the boundary-time" (Jones 1978: 229) some 25 to 30,000 or more years ago?

Modern can, as in "Modern Man," mean the "viable past." I can, for example, when I am reading David Hacket Fischer's *Paul Revere's Ride*, find myself out for a stroll in the eighteenth-century wing of in-my-time. During the stroll I realize that, unlike many of my contemporaries in the 1990s who believe that the right to free speech is individual and personal as in "It's my right," I still harbor the eighteenth-century belief that rights are collective and that responsibility is individual. If freedom of speech is a collective right, my right to speak my mind has to be governed by the ethical provisos of my responsibility, provisos which include careful attention to the perceived truth and due consideration of context and the probable impact of what I am saying. The stroll in the eighteenth-century wing of "modern" was an exercise in incorporation of the viable past, an exercise in *paideia*.

In another wing of the modern memory palace, John E. Pfeiffer's book *The Creative Explosion* can take me for a walk in the dim light of the Boundary Time when my contemporaries, Early Modern Man, were preparing to undertake the murals at Lascaux and at Altamira. The priorities of our practical materialism lead us to focus on the explosive development of the Cro-Magnon tool kit. Those are the same priorities that lead us to assume that language arose out of practical necessity and only latterly developed the capacity for metaphor and symbol that floats the sign-stream of culture and civilization. Those priorities marginalize(d) the ur-art of memory and the melody of consciousness and action. But I like to think that Whitman was right about body and soul "and the unseen is proved by the seen" (Whitman 1982: 29). Another exercise in incorporation, in *paideia*.

But the "viable past" must extend beyond the constraints of my memory or of any one of our memories into the realms of, in David Jones's phrase, "unshared backgrounds." We do not all incorporate and share the same aspects of the viable past, and experts among us can disagree, often violently, about the aspects we should share. The extent to which we do not share can be brought home to us when we step into some of the major literary works of our time such as T. S. Eliot's *The Waste Land* or James Joyce's *Ulysses* or David Jones's *Anathemata* and find ourselves in need of a bewilder of footnotes. We may be tempted to ask: why don't these twentieth century contemporaries of ours tell us in plain words the way the Old Masters did? And then we can be interrupted by the reminder (and it should be a rude reminder) that we need a similar bewilder of footnotes if we are to find our way around in Dante or Shakespeare or Milton, or even in Homer and the Bible, for that matter.

In the lifelong mnemonic enterprise of *paideia*, we move from the private layerings of our autobiographies into the larger realms of family and village and the more formal communities and cultures beyond. David Jones opens the "Preface" to *The Anathemata* with a line from Nennius, the early ninth-century Welsh monk whom we suppose to have been the compositor of the *Historia Brittonum*: "I have," Nennius says, "made a heap of all I could find" (quoted in Jones 1978: 9). And Jones announces that he, too, is making a heap out of all the Anglo-Welsh and Roman Catholic backgrounds he can find. Jones calls the complex resources of the viable past "the deposits," and continues:

> Clearly they comprise in our tradition a great complex of influences and interactions which have conditioned us all. To say that one draws upon such deposits does not imply erudition; it suggests only that these form the *materia* that we *all* draw upon, whether we know it or not, to this degree or that, in however roundabout a way, whether we are lettered or illiterate, Christian or post-Christian or anti-Christian. Then there are the living, dying, or dead traditions in which one has oneself participated or heard of with one's own ears from one's own parents or near relatives or immediate forebears. These things received in childhood are of course fragments or concomitants only of the whole above-mentioned complex. I am thinking only of the means whereby these concomitants and fragments reached me. I am speaking of channels only, but of immediate channels and such as condition all that passes through them, and which condition also one's subsequent attitude to all the rest. These I judge to be of the most primary importance. It is through them that 'all the rest' is already half sensed long before it is known. If ever it is known. (Jones 1978: 40-41)

And he concludes: "You smell a rat or two pretty early on" (41) in the gap between the knowing and the not-ever-knowing where we all, more or less, reside.

Chapter 5
The Sign-Stream of Our Histories

Prologue

Briefly to recapitulate the interrelation of language and memory (human memory): the relation is not causal; language and the capacity for narrative do not cause memory. They are interrelated as a mix of preconditions. Memory is a precondition of language; and language is the precondition of human memory. This is not to say that memory is verbal or that it is all narrative. We have memories of visual images, tactile images, of sound and odor and taste; and we have kinetic memories (how to walk, sit down, stand up, ride a bicycle, and so forth.). The point is that we can story those memories because we have language and narrative. We don't have to story them to remember them, but they are available to language if and when we so please. The primary function of language as we enter it may well be the brain-making it enables and which accompanies our immersion in it. Language and memory together are the preconditions for the self-making and world-making which is our individual and collective enterprise as human beings.

1. The Sign-Stream of Our History

As a point of departure, a Shaker proverb: "History is to the community what memory is to the individual." As memory is to consciousness, time, individual identity; so history is to the community's "stored narrative resource." It harbors what David Jones calls the "deposits," the "sign-stream" that flows toward and through us from the "viable past." And in the sign-stream of our histories there is a powerful emphasis on commemorative days, on ceremonies of remembrance. These occasions mark the cycles of the week and the month and the year and the cycles of our lives from birth to maturity, to marriage and family and children and death in the religious, national, and smaller communal histories that dominate the deposits and the sign-streams of our world. And we participate in these markers individually as well as collectively.

Thoreau moved into his cabin on Walden Pond on the Fourth of July, Independence Day, 1845, marking an intersect between a personal assertion of independence and the national celebration which in 1845 would have been marked not only by parades but also by real speeches, not five minute sound bites but two and three hour orations, the great national popular entertainment, one act of which was going on to the tune of fife and drum in Concord while Thoreau sought silence and sanctuary for "some business" he wanted "to transact" in the second growth woods around Walden Pond a mile and a half away. Nine years later that private assertion of independence would become a dramatic public assertion when the book *Walden* quietly slipped into the sign-stream of the American mind. And I do mean "quietly." The noise the book was to make was at least half a century (more like a century) away when the book was first published.

In the great religious sign-streams of our heritage, Pesach (Passover) celebrates not only the season of the liberation of the Children of Israel from the House of Bondage in Egypt, but also (in a broader sense) the beginning of that people's journey toward spiritual identity. Easter, which occurs during the same season of the year, marks the annual beginning, the baptism, of the Christian Liturgical Year. In the Gospel account what will become the Easter Weekend begins when the Disciples ask Jesus "Where wilt thou that we prepare for thee to eat the Passover?" (Matthew 26:17), and he answers by giving them instructions for the setting of what we will in retrospect call the Last Supper. That meal will center on the bread and wine and the instruction, "This do in remembrance of me" (Luke 22:19).

Whatever our religious orientation, many currents in the sign-stream encourage us toward seasonal remembrance and renewal. The coincidence of Pesach (Passover) and Easter has by some (among them Sir Thomas More) been strenuously denied in the course of an elaborate and centuries-long controversy. The end result is the complex gerrymandering of the Christian calendar that keeps the two festivals from coinciding more than three or four times a century. But I'm happy to announce that the purity thus assured to Easter is somewhat undercut by the name Easter itself which derives from Old Teutonic mythology, from "*Eostre*, or *Ostâra*, the Anglo-Saxon goddess of Spring, to whom the month answering to our month April, and called *Eostur-monath*, was dedicated," Easter, when, as the English teacher, historian and theologian, the Venerable Bede (673-735) said "the old festival was observed with the gladness of a new solemnity" (*Britannica* 1910-1911: VIII:828).

G. K. Chesterton once said that coincidence is a sort of divine pun, and the coincidence of Pesach (Passover) and Easter, while denied by those Robert Graves calls "the prose men" who jigger calendars, has had a vigorous life in the sign-stream of our viable past. The Greek name for Easter we would translate as Passover, and both the Haggadah of Passover and the Easter celebration in the Eastern Orthodox Church pivot on Psalm 114:1-2 (Vulgate, 113) as does the litany of Easter Vespers in the Western Church:

> When Israel went out of Egypt, the house of Jacob from a people of strange language;
> Judah was his sanctuary and Israel his dominion.

In the west that psalm was to become central not only in Easter Vespers and in Sunday Vespers throughout the year but also in Dante's *Divina Commedia*. That epic takes place during the idealized Easter Season of 1300, the year Dante (1265-1321) was at thirty-five in "the middle of the way of this life," which was according to Isaiah, scheduled for "three score and ten." In the *Purgatorio*, Canto II, lines 46ff, the Psalm of Exodus is being sung by the souls of the fortunate dead when Dante and Virgil see them being ferried to the foot of the mountain of purification by Charon's angelic counterpart. These dead are "fortunate" because, unlike the souls Dante has encountered in the *Inferno*, they have the hope and the assurance of eventually entering paradise. The Psalm of Exodus is also the psalm that Dante used in the letter to his patron Can Grande della Scala, dedicating the *Paradiso* to him. Dante quotes the sentence I have just quoted and then remarks,

> Now if we look at the letter alone [the literal sense], what is signified to us is the departure of the sons of Israel from Egypt during the time of Moses; if at the allegory . . . our redemption through Christ; if the moral, . . . the conversion of the soul from the sorrow and misery of sin to the state of grace; if . . . the anagogical, . . . the departure of the sanctified soul from bondage to the corruption of this world into the freedom of eternal glory. (Dante 1973: 99)

As the Psalm of Exodus is pivotal in the transition from the terminal depression of hell toward the hopeful looking-forward of purgatory in Dante's epic, so also in Joyce's epic of the middle-class way of our lives, *Ulysses*. Just short of the end of that novel, the two heroes, Stephen Dedalus and Leopold Bloom, have cups of a Last Supper cocoa together in the basement kitchen of Bloom's house. They then exit into Bloom's back garden under a star-rich sky (as Dante and Virgil have crawled out into the pure night air from the oppressive tunnel that leads from the bottom of hell at the earth's center to the foot of the mountain of purgatory opposite Jerusalem). As they leave the house Stephen chants the Psalm of Exodus, and we realize that Bloom, the lapsed Jew, is playing the part of Virgil and that Stephen, the lapsed Catholic, is playing the part of Dante.

When, a bit later, Stephen leaves Bloom's back garden he also departs the novel. Bloom, in the middle of the way of this life, is left behind in the middle way in which we assume he will remain as Virgil (in spite of all his virtues) is destined to remain in the "hemisphere of light" among the virtuous heathen in Limbo. In what direction Stephen's steps will take him we cannot be sure though the powerful undercurrent of the sign-stream encourages us to hope that Stephen might move from the paralytic doldrums of the past year of his life (as that year has been unfolded for us in the retrospective course of the novel) in the direction of some sort of purgatorial experience, some sort of liberation which might enable him to pass over into the poet, the Dante, he aspires to be.

And I would argue that the elaborate efforts (such as Sir Thomas More's) to cordon off Easter from Passover and to prove that the Last Supper was not a Seder threatens to diminish the power of the sign-stream rather than to assure its orthodox purity because the power of that coincidence can survive even the transition from fourteenth century Christian epic to inform a twentieth-century secular epic and thus can doubly inform us.

Back to the Shaker proverb: history is to the community what memory is to the individual. I think of history in terms of two quite different paradigms or models, one of which I call centralist history, the other fractal history. Since these are my terms, I had better define them. By centralist history I mean history organized around a central core of cause, either secular cause as in the Course of Empire or Manifest Destiny or divine cause as in the course of orthodox Christian history: from the fixed beginning in the Garden of Eden to the fixed middle on the Cross to the fixed end, Armageddon and the Last Judgment. That is radically centralist as against the muted centralist histories which trace the cause and effect patterns that purport to explain the so-called "victory" of capitalist democracy over communist totalitarianism.

Fractal history derives its metaphor from fractal geometry, which came our way when one twentieth-century mathematician suggested that "Nature has played a joke on the mathematicians. The nineteenth-century mathematicians [who were tied to Euclidian geometry and Newtonian dynamics] may have been lacking in inspiration, but Nature was not." Benoit B. Mandelbrot, who invented the word fractal, illustrates this point by asking "How Long Is the Coast of Britain?" He observes that "the typical coastline is irregular and winding, and there is no question it is much longer than the straight line between its end points"

(Ferris 1993: 449). He suggests that if we set dividers to a yard and then walk them along the coast, and then refine the measure to a foot, to an inch, and then to fractions of an inch "we expect [the measurement of length] to settle rapidly to a well defined value called the *true length*. But in fact what we expect does not happen. In the typical case, the observed [length] tends to increase without limit" (Ferris 1993: 450).

Similarly with history: we expect the "story" to settle gradually toward the "true story" which will (we Hope and Trust) be established by just one more review when all the facts are in and it can be settled forever. But history keeps coiling around us in all its multi-directional complexity, threatening even the stability of the present tense in which we are condemned to write it, the present tense in which via history we affirm the mnemonic vitality and diversity of community. With the reminder that history is not comfortably back there in the past, over and done with, just as the future is not comfortably out there, waiting to happen because in what Robert Graves has called those "tourbillions [whirlwinds] in time" the future may already have happened. Fractal history: we cannot know, cannot trace all its courses, tributaries and eddys and backwaters until after the end of history when, of course, we won't be around to do the tracing and the summing up; and if there is to be a final summary, I cannot imagine it in other than fractal time, its web of happening tending to increase without measure.

In contrast to fractal history centralist history tends to assume a past centrality of purpose (divine or human) which, despite the distracting multiversity that hides it from us, will emerge triumphant at some future rendezvous with time. What made Charles Darwin's theory of evolution so profoundly disturbing to his contemporaries, what made it an agony for Darwin himself, was that its fractal complexity so decisively challenged the comfortable Victorian version of the faith that "[a]ll human history moves toward one great goal, the manifestation of God" (Joyce 1986: 2.380-81). The Victorians wanted to see direct evidence of that progress in the transformation of the world of space and time being achieved by the industrial revolution which transformation they took to be evidence of the moral and spiritual progress of mankind.

Darwin's theory challenged all of that not only because it seemed to deflate human dignity by saying that man was descended from the apes instead of from angels, but more profoundly because it said in no uncertain terms that history was fractal rather than purposive; that history was a concatenation of random events rather than a working of underlying and purposive cause, however mysterious that working might appear to those of us confined to the surface. That desire to believe in underlying and purposive cause was and has remained so powerful that ostensible believers in Darwin's theory have given evolution itself a goal. In that centralist reading, the purpose of evolution is progress toward man and human consciousness, and the further purpose of evolution is progress toward higher and higher levels of moral and spiritual awareness. So the impact on human history of what Darwin had to say about biological history has been obscured by the imperatives of centralist history. And when biologists say about one extinction or another "that particular evolutionary experiment failed," the language itself betrays them into an unintended centralism because to experiment is, we assume, to experiment to some purpose known to the resident intelligence that designed the experiment in the first place.

As Stephen Jay Gould, a proponent of what I would call fractal history, puts it: "we yearn to depict our complex historical present as a necessary outcome of former conditions, and not as an unpredictable, if sensible, result that would never occur again if we could replay life's tape from an identical starting point" (Gould 1995: C5). In other words, while we live in fractal history, we yearn for the securities of some sort of centralist history.

I've wondered for some time (and still wonder) how centralist history, faith in some sort of progress on the wings of our gee whiz technologies, some alleviation if not outright improvement in the human condition, has survived into our own time. Shouldn't it have been terminally interrupted by the impressive demonstration of what the mechanization of massacre could accomplish on the Somme and at Verdun in 1916? But not at all. We are survivors of a century of massacre as one parenthesis of massacre has followed another. First there was the parenthesis (1914-1918) that the Anglo-Welsh poet-artist David Jones has focused so splendidly in his prose poem *In Parenthesis* (1937). That parenthesis was followed by what looked like another to David Jones and those living through it, the parenthesis of peace, 1918-1939 between two bouts of the real thing, 1914-1918, 1939-1945. And most recently we have survived what I call the MAD Parenthesis of the Cold War (1948-1991). MAD, after the initial letters of the strategy that both sides shared: Mutual Assured Destruction—that is to say, at least mutual, if not global suicide.

I want to pause here to consider the complex relationships between a history and the community which it shapes and informs—as the self is shaped and informed by autobiographical memory, by narrative truth rather than by strict historical truth, so the community. If we accept Emerson's rhetorical question, "What is history but a fable agreed upon?" then we might go on to raise the question: what are the ethical and aesthetic values offered or compelled by that agreed upon fable? Ethical in the sense: how does that fable address the question: what is the good life, a life well-lived and what are the moral disciplines necessary to its fulfillment? Aesthetic in the sense of how do the hushed reverbrations and retrievements of that agreed upon fable enrich the lives of the communities that share it?

And so back to the MAD Parenthesis and its presence in our community, the world community. Whether we agreed with the MAD centralist *realpolitik* fable upon which it was based or were in fractal but loyal dissent does not matter here. That parenthesis was a presence in our lives, individually and collectively, never mind the degree to which the thought of Nuclear Holocaust compelled, in Doctor Johnson's phrase, "the mind to relieve its distress in incredulity" (Baym 1989: I:2360). Because what that MAD Parenthesis asked us to contemplate was the annihilation of most of humanity, and, for those who survived, the devastating loss of any sense of humanity in the medical vacuum, no treatment available for the multitudes of the injured; no one governable and no government left unless you count some committee airborne above the radioactive dust or buried incommunicado (no communication, no community) beneath a mountain somewhere on what might as well be Mars. And there was always the subsidiary strategy, second fiddle to Mutual Assured Destruction, of "decapitation"; not "assassination," mind you, but just take out the other side's "command and control centers."

This MAD situation excited e. e. cummings to vicious irony: the United States is a nation, he said, "dedicated to the proposition that massacre is a social virtue because murder is an individual vice." Lewis Thomas, in a more serious vein in *Late Night Thoughts on Listening to Mahler's Ninth Symphony* was of the opinion that the subliminal infection of the MAD Parenthesis would have driven him "crazy" were he sixteen (Thomas 1983: 166). And that raises the disturbing and unanswerable question: to what extent has that fable, agreed upon among those so remote from us that we can only imagine them as unmoved movers, to what extent has that fable driven all of us crazy, into seizures of impotent rage, into a destabilization of ethical values, into a terminal cynicism: if a purposeless Armageddon is to prevail, what does anything I do or can do matter?

And when the Cold War ended, close parenthesis, what could be sillier than to celebrate that close with a book entitled *The End of History and the Last Man* as the conservative think-tanker Francis Fukuyama did? And to celebrate it by saying: "In the post-historical period there will be neither art nor philosophy, just the perpetual care-taking of the museum of human history" (Fukuyama 1989: 18). The Last Man, indeed, when we are not clearly Post-Modern let alone Post-Historical and when it is all too clear that the infection we call history lingers on in ways we still cannot satisfactorily diagnose, let alone understand.

I must admit that just spelling that MAD parenthesis out for the retrospect of this lecture has been an exercise in delayed anxiety, but it has set me to wondering what a benign agreed-upon fable would look like. There are plenty of examples of histories that have had and continue to have more or less malign effects. But benign? That Shaker proverb comes again to mind, and with it, the fable the Shakers were agreed upon. Much of Shaker doctrine has been attributed to Mother Ann Lee (1736-1784), but it continued to be elaborated well into the nineteenth century. The Shakers divided history into four "dispensations." The First Dispensation or Cycle, from Adam to Abraham, was the period of Darkness after the Fall of Man. The Second Dispensation began with the revelation to Abraham of the Lord's "Covenant"—that Abraham's "seed" would be a "chosen people" in the "Promised Land" of Canaan (Genesis, chapters 12-25). The Third Dispensation was announced by the appearance of the male Messiah, "For as in Adam all die, even so in Christ shall all be made alive" (I Corinthians 15:22). But in the Third Dispensation back-sliding and unregenerate man realized not the Kingdom of Christ but the Kingdom of the Antichrist. The Fourth Dispensation, the Millennium, was to be realized at the Second Coming, and that had already happened in the person of Mother Ann whose mission it had been to complete the revelation of God to man.

But if we step back and put that doctrine in the fractal context of late eighteenth-century rural America, it assumes a different shape. In that context there was considerable religious ferment, revivals, and awakenings in isolated rural communities, sparked by the expectation that the Millennium was near at hand, due, for example, a week from Sunday next. And when the due date had passed on its fractal way, the fervor that had transformed the community often gave way to agonies of depression. That had apparently been the case in New Lebanon, New York, late in 1779. And when two of the depressed revivalists visited Mother Ann in Niskayuna early in 1780, their spirits were revived by the news that the Millennium had already come. The result was the rapid conversion of other New Lebanon revivalists and the establishment of the closed communal structure of the Millennial Church of Zion on

Earth. In short, the Shakers undertook to sustain the benign fervor of revival by preaching that the Millennium had already happened and that they were living in it, beyond the interface that separates life from the hereafter. They were living in what they called "the regenerate order" while non-Shakers, "the World's People" or "Adam's Kind," lived in "the realm of generation," the hell that was life-on-earth for those yet to be converted and welcomed to the Covenant.

In the fractal history of the Shaker communities life went its usual way of spiritual and human ups and downs. The achievements of the practical economies of the communities was impressive, and as the communities aged, their neighbors came to regard them not as hives of dangerous radicals but as the retreats of sober, quiet-spoken, and dependable (if eccentric) neighbors. Fractal history has overtaken the Shaker communities one by one as their efforts to make new converts have come up empty, and the heavenly enclaves where "they neither marry, nor are given in marriage" (Matthew 22:30) are all but empty.

William James once suggested a way to measure and assess the benignity of a centralist history. In 1898 he delivered and published a lecture, "Human Immortality: Two Supposed Objections to the Doctrine." It was the second annual lecture in a series, which is still continuing at Harvard, the Ingersoll lectures on the Immortality of Man. In the historical context in which the lecture was delivered "immortality" meant "the endless existence of the human soul *in the continued possession of its distinct personality and consciousness*" (Schmidt 1913: 496). The establishment of the Ingersol lectureship at Harvard in 1897 reflected a popular and academic interest in discussions of the subject, an interest that found in "immortality" a sharply defined arena where the conflicting claims of the new sciences and the old faiths could meet metaphorically in single combat. The question of immortality also seemed to hold out a promise of relief to those worried by the late nineteenth-century sense that man had been reduced from his former universal significance to universal insignificance by the radical alterations in space (from the intimate spaces of earth to inter-galactic space) and in time (from the comfortable six thousand or so years since Creation and Eden to eons of geological and biological time). This sense of vertiginous change was also re-enforced by the transformation of experiential space-time being achieved by continuous mechanical motion in travel and by instantaneous electronic communication. The world was no longer, as William James put it in his lecture, "a comparatively snug affair" (James, William 1988: 31).

James initially said that he wasn't much interested in immortality as a topic, but then he thought better of it, thought that it might be fun to knock the props out from under those he called the "physiological psychologists" (the radical materialists) and their supposedly scientific objections to the doctrine of human immortality. The demolition was accomplished in the opening phases of the lecture, and then James turned to the audience and said, "you *may* believe." But to that rather tentative permission he added a qualification: provided it improves your sense of "the inner significance of other lives" (James, William 1898: 44), a significance which, he remarked, "exceeds all our powers of sympathy and insight" (James, William 1988: 44). He observed in passing that the historical track record of those who were zealous believers in the doctrine of human immortality frequently demonstrates the polar opposite of anything that could be called sympathy and insight.

The fractal histories available to us provide overwhelmingly copious evidence. The Crusades to liberate the Holy Land from the Muslim infidels were preached with clear centralist goals, but they were undertaken in the fractal realms of Medieval warfare: massacre, carnage, and plunder. The inside-Europe Crusade against the Albigensians in the eleventh and twelfth centuries was supposed to ensure the purity of the Holy Roman Apostalic Church. The Inquisition torched the Albigensians in a triumph of purity, with the incidental side effects of demolishing the thriving economy and the rich culture of the Languedoc region in the south of France. In another impressive demonstration of agreed-upon fable, the Franciscans came down with considerable severity on the Aztecs in Mexico because the Franciscans believed that legendary travelling saints like the Irish Saint, Brendan the Voyager, had reached and converted the New World long before Columbus, and in their eyes Aztec iconography bore enough resemblance to Christian iconography to "prove" that the Aztecs were lapsed Christians who had earned the rod and the scourge, not heathens who might have deserved a drop or two of the milk of human kindness.

And when the religious wars of the seventeenth century were over, we had the emergence of the modern nation-states with their secular, but still impatiently centralist enterprises: the course of empire, the white man's burden, manifest destiny, right down to *lebensraum* in World War II and the now of the Balkans, a region which, as Churchill once said, has produced "more history than it can consume."

By a not very circuitous route we come home to our own century of massacre and the ideological states which have provided the centralist histories and theories that have achieved the wholesale and global destruction of communities and their histories, their narrative resources. The Soviet Union undertook to level down the communities of its empire to concrete block homogenization; in the process churches and religions were swept aside with no credible replacement, and thousands of peasants and their communities bull-dozed in the interest of collectivization. The Chinese occupation of Tibet proceeds apace to demolish the historic buildings of Lhasa, together with thousands of homes in order to erase the historical memory of Tibet and to make Lhasa "a Chinese city" (Schell 1994: 12). The Chinese, as Jonathan Schell remarked, "want the soul of Tibet" in an exercise of historical cannibalism (Schell 1994: 12). The Khymer Rouge once proposed to level Ankor Vat, to edit it out of the history of Cambodia, so that Cambodia could become, the ideological ideal, a rootless homogenized state. And Sheik Abdul Rakhman has advocated the demolition of all the pre-Islamic monuments of ancient Egypt. That would, of course, purify Egypt for Islam and keep the tourists from contaminating the country into the bargain.

These scorch-the-earth policies, these zealous attempts to cut off the flow of the sign-streams of history are all around us in the world of the twentieth century, and nowhere more devastatingly and climactically than in Nazi Germany's determination to exterminate European Jewry. We think of that campaign in terms of *einsatzgruppen* and mass graves, of gas chambers and crematoria, but it was much broader and more sweeping than just death. It was a campaign to humiliate Jews individually (to fracture the self and the autobiographical memory that sustains it), to break up families and their narrative resources, to fracture community by squeezing the population into the pressure cookers of ghettos and concentration camps, to wipe out the community by wiping out its history. It was a massive exercise in what Erik Erikson has called pseudospeciation, turning another group of human beings into an alien species in order to make it psychologically and emotionally possible to despise them, to hate them, to kill them.

And in all of this, a grisly paradox: Hitler's determination to establish what was to be called "The Museum of an Extinct Race" in Prague, his determination to wipe the Jews out of history and then to remember them in a museum. To that end the Nazis collected vast amounts of material at what seems to have been almost mindless random, black comic random as that is portrayed in the Czech writer Jiří Weil's novel, *Mendelssohn Is on the Roof* (1991). The Nazis also collected silver and other valuable artifacts from synagogues throughout the territories they had overrun. Some of those treasures are now on exhibit in the Museum of Jewish Silver in the former Maisl Synagogue in Prague. That museum and the Museum of Jewish Life nearby in the old ghetto are strange experiences: so serene and pure, as though they just dropped in from eternity, when in memory there is so much darkness and distress, so much fragility and mortality. The unanswerable question lingers (and with extraordinary urgency): how to remember; how to commemorate?

That urgency rings like a clarion call through Israel Lichtenstein's "Last Testament," written in the Warsaw Ghetto, 31 July 1942, nine months before the Warsaw Ghetto Uprising and the subsequent massacre in the Passover season of April 1943. Lichtenstein had been appointed the secret archivist of the Ghetto; he was placed at the heart of the Ghetto's memory web:

> With zeal and zest I threw myself into the work to help assemble archive materials. I was entrusted to be the custodian. I hid the material. Besides me, no one knew. I confided only in my friend Hersh Wasser, my superior.
>
> It is well hidden. Please God that it be preserved. That will be the finest and best we achieved in the present gruesome time.
>
> I know that we will not endure. To survive and remain alive [after] such horrible murders and massacres is impossible. Therefore I write this testament of mine. Perhaps I am not worthy of being remembered, but just for my grit in working with the society Oneg Shabbat and for being the most endangered because I hid the entire material. It would be a small thing to give my own head. I risk the head of my dear wife Gele Seckstein and my treasure, my little daughter, Margalit.
>
> I don't want any gratitude, any monument, any praise. I want only a remembrance . . . I want my wife to be remembered. Gele Seckstein, artist, dozens of works, talented, didn't manage to exhibit, did not show in public. . . . I want my little daughter to be remembered. Margalit, 20 months old today. . . I am not sorry about my life and that of my wife. But I am sorry for the gifted little girl. She deserves to be remembered also. (Quoted in Dawidowicz 1996:296-97.)

As I read that over again, I am struck by the full urgency of the need to remember, to commemorate, and the always open, all but unanswerable question: but how? All the usual ways—museums, monuments, annual Days of Remembrance—seem uncertain, vulnerable to co-optation by politics, by a sort of competition for possession rather than remembrance. God forbid that Auschwitz, sterilized and all tidied up, should become a theme park. Isn't it enough just to say that Auschwitz had the largest railroad yard in Eastern Europe?

The Viennese philosopher Sir Karl Popper (b. 1902) has addressed one slant of this question with, according to one commentator,

> the penetrating insight that ignorance is not a simple lack of knowledge but an active aversion to knowledge, the refusal to know, issuing from cowardice, pride or laziness of mind. Popper did not accept the explanation that people did not know about the concentration camps, the Gulag, the Holocaust. They simply did not want to know. In Popper's philosophy, ignorance has an ethical dimension, and knowing is a moral obligation for human beings. (Ryszard Kapuscinski quoted in Dawidowicz 1996)

In short, not knowing, not remembering (let alone active forgetting) is evil. And when Albert Speer said to his biographer Gitta Sereny "I was blind by choice . . . but I was not ignorant" (quoted in Sereny 1995: 222), the hair he was splitting about his knowledge of slave-labor conditions in his factories and about the Holocaust was the hair of evil; split one evil, result: two in one.

Jerome Bruner identifies another comprehensive threat of break-down in culture and its narrative resources as inhering

> in the rhetorical overspecialization of narrative, when stories become so ideologically or self-servingly motivated that distrust displaces interpretation, and "what happened" is discounted as fabrication. On the large scale, this is what happens under a totalitarian regime, and contemporary novelists of Central Europe have documented it with painful exquisiteness–Milan Kundera, Danilo Kis, and many others. The same phenomenon expresses itself in modern bureaucracy, where all except the official story of what is happening is silenced or stonewalled. And finally, there is breakdown that results from sheer impoverishment of narrative resources—in the permanent underclass of the urban ghetto, in the second and third generation of the Palestinian refugee compound, in the hunger-preoccupied villages of semi-permanently drought-stricken . . . sub-Saharan Africa. It is not that there is a total loss in putting story form to experience, but that the "worst scenario" story comes so to dominate daily life that variation seems no longer to be possible. (Bruner 1990: 96-97)

No wonder we so continue to seek benign communities and the nourishment of their narrative resources (as on these afternoons in this room).

Chapter 6
Memory, the Self, and Art

At the end of the last lecture I quoted Jerome Bruner at length about assaults on a community's narrative resources, assaults on communal memory that threaten breakdown in a culture. Bruner cited, among others the assault behavior of totalitarian regimes and the plight of "the permanent underclass in the urban ghetto." We are all appropriately appalled when we read that Nicolai Ceauceşcu, the dictator of Rumania, organized the orphanages of that unfortunate nation as training camps for the brutalization and dehumanization of children destined to become recruits for his Securitate. And we should be similarly appalled when we read the comments of an eye-witness to the drug wars in Queens, New York: "Most of them [the drug warriors] raise themselves on the street. And they are raised hard. They have no values in life." Or as an older woman in a Chicago Housing Project said of her drug-warrior neighbors, "Those people don't do no talking." Heirs to an impoverished narrative resource, they have been all but deprived of the grammar of anticipation. They are reduced to a grammar of contingency, reacting to the now of whatever is happening in an environment overwhelmingly dominated by violence, chance, and accident.

And so to the final lecture which is really two: one on memory and the self, the other on art as a way of knowing, art as a fundamental and much neglected way of nourishing memory and consciousness.

1. Memory and the Self

Memory and the self: my desk dictionary defines self as "1. The total, essential, or particular being of one person, the individual. 2. The qualities of one person distinguishing him from another . . . 3. An individual's consciousness of his own being or identity; subjectivity; ego." The first and third definitions suggest an emphasis on the view from within, from behind the eyes out toward the world. The second and (to a lesser extent) the first suggest a view from outside, distinguishing one person from another, as I can tell you from you, and if I step outside myself, you from me. The view from within is unselfconsciously ours in the day-to-day of our experience, the view from outside is available in our occasional (and usually self-conscious) up by-the-bootstraps efforts to see ourselves as others see us.

The view from within contemplates a self that is variously and elusively centered, a self that, in the course of a day, is involved in many roles; there are, in effect, many selves, all of them sustained and timed and fine-tuned by memory. There is the self in the roles prescribed by the day's routines: shave and dress (whose face is that?); grind the coffee, lay out breakfast. There is the role of husband and companion, the amusing alarm clock. Another role is added with the retrieval of the morning newspaper and the complex mnemonic processes that sort, inform, and highlight its mix of news and comics and irritations. There are handyman roles after breakfast before the role of the chap who sits at this keyboard preparing a lecture for a role and for a voice quite different from the one he used at the breakfast table. And beyond the intimate circumference of the home there are other selves: those that think about and talk to scattered members of the family, those that deal with friends and acquaintances, those that do the errands in the village and beyond. And throughout, memory is there with its clutter of maps and timetables, facebooks and voicebooks and sometimes even names to go with them, memory with its fabulous support systems, including language and the voices appropriate (we hope) to the range of the day's occasions. To bend Shakespeare's Richard II, "Thus play I in one person many people / And [some] of them contented" (Shakespeare V:v:31-32).

And from a neurological point of view, this multiplicity of selves is apparently a good thing. Israel Rosenfield remarks that "the senses of self and speech, like posture, are constantly evolving structures"; and he says about multiple personality disorder that the "patient . . . has too few selves . . . Pathology has limited the responses and therefore the dynamic, ever-changing nature of self" (Rosenfield 1992: 122-23).

Emily Dickinson once remarked (in a letter to T. W. Higginson, July 1862) "My Business is Circumference" (Dickinson 1992: 176). And by that she apparently meant that it was her business to project "her imagination into all [that is, as many] relationships of man, nature, and spirit" as she possibly could (Baym 1989: I:2406, n3). And she jealously patrolled her Circumferences with considerable apprehension lest her sanctuary within be invaded by a world that she probably quite correctly perceived as hostile to her private strains of non-conformity and heresy. Most of us tend to react that way, hardening our Circumferences when we think we are treading on polite but unfriendly ground, a sort of ground which in her case extended at least as far as the front steps of her house.

What St. Augustine said of memory and the mind can be said of the self and its circumference as well, "the mind is not large enough to contain itself," and when he asks the three still unanswerable questions: "But where can that uncontained part of it be? Is it outside itself and not inside? In that case, how can it fail to contain itself?" (Augustine 1963: X.8, 219). One way to approach an answer to those questions may be to suggest that just as the self is elusively multi-centered, so, too, is memory with its circumferences and the riverine courses that support and inform our perceptual attempts to explore them. When I sit here at this desk, for example, at the heart of a cluttered web of note cards, journals and file folders full of notes and lectures and other manuscripts, and all the stuff I've squirreled in that computer and on those diskettes, and all the books beyond on the shelves, I imagine all of that as somehow within the shelter of my memory, part of myself, even "if it is not contained within it."

I don't think of the self and its selves as things that exist by virtue of verbalization, save perhaps temporarily (and one hopes privately) in the course of psychotherapy, and certainly not in the psychobabble confessionals of popular culture, its afternoon talk shows and its bulletin boards on the internet. In other words, I don't think I have a self and its selves because I can talk it/them, but because I am them and act/behave them.

And when people sum up their lives, their several selves, verbally, we prize an elegance of indirection whether in well-written autobiography where the adjective well-written implies, as Robert Lowell implied when he said of his so-called "confessional poems," they are "full of lies." And there is the indirection of sustained metaphor. At the end of his life when he was laid up in a Dublin nursing home, physically no longer able to paint, the outstanding Irish painter Jack B. Yeats (1871-1957, William Butler Yeats's younger brother) amused himself by composing various versions of his "last words." As a painter he had been regarded as a radical who could not paint; that is, he was out of step with his times or rather out of step with Ireland and its times, not a very difficult thing to be in that culturally conservative or rather culturally reactionary country in the opening decades of this century.

> I have travelled all my life without a ticket: and, therefore, I was never to be seen when Inspectors came round because I was under the seats. It was rather dusty, but I used to get the sun on the floor sometimes. When we are asked about it all in the end, we who travel without tickets, we can say with that vanity which takes the place of self-confidence: even though we went without tickets we were never commuters. (Quoted in White 1957: 121)

I should add that the Irish despise commuters (as they despise civil servants) because commuters and civil servants are aping British manners; they are being West Britons, lackeys of the British.

When we observe and contemplate the impact of memory-loss on brain-damaged people, we begin to realize how impossible it is adequately to verbalize the self and its selves and the flows of memory that keep it/them afloat. Oliver Sack's account of the patient he calls "The Lost Mariner" makes it clear that memory is at the heart of self. The patient, he says, "had no 'day before'" (Sacks 1985: 34), no antecedents which could nourish the connective tissue of memory; he has lost the memory-self that would know he has lost the self that once was his.

As we have watched friends and acquaintances descend into Alzheimer's Disease, we watch memory holes open. One close friend, a jazz pianist and composer, has lost the ability to play the piano though his tennis skills remain relatively intact. At the keyboard memory and anticipation of the next improvisatory phrase are no more, but on the amateur tennis court where contingency reigns he can still function. His ability to understand his wife's absence is unrelieved by the mnemonic ability to anticipate her return. The awful part of this is that the distress is there but the flow of memory that would relieve it has been permanently interrupted, and other memory flows are disappearing into black holes as time gathers him in. In another case, a remote acquaintance who had been temporarily hospitalized because her husband had been slowed by a stroke, escaped in her distress from the hospital and walked home, almost six miles across central London, without accident but with no anticipation or apparent awareness of what the physical toll on unshod feet would be. Another acquaintance sat comfortably chatting with four people, two of whom she knew. She left the room, and when she returned a few minutes later, she sat down quietly and then suddenly demanded (with considerable anger) to be introduced to us, the two strangers.

I bring up these random examples to suggest how dependent the self and its selves are on the nourishing flow of memory. We're all aware of how disturbing momentary lapses of memory can be, but for most of us those lapses are knowable and at least reparable (or at least forgivable), but if not . . .?

In the fifth lecture (The Sign-stream of Our History) I spent some time considering the ways in which assaults on history, assaults on the narrative resources of a community, have been mounted by dedicated believers in rigidified centralist histories and by zealous ideologues. I want to return to that terrain but from the perspective of the individual self and the yet-to-be-completed rough draft of its personal autobiography. One way to explore the circumference of the dark side of memory and the self is through the testimony of people who have been tortured, people who have suffered the trauma of memory-loss through the trauma of torture or who suffer the sustained trauma of being unable to forget the nightmare experience—as in the terrifying seventeenth-century formulation, "One torture will follow another until you say you are a witch."

That exploration might take us to Primo Levi's (1919-1987) *Survival in Auschwitz*, with its subtitle: *The Nazi Assault on Humanity*. When it was first published in 1959 many readers were led to hope that the writing of the book would be for Levi an act or ceremony of purgation which would free him of the nightmare. And when *Survival in Auschwitz* was followed fifteen years later by another memoir, *The Periodic Table*, that hope was renewed, not least by the Yiddish proverb that was its epigraph: "Troubles overcome are good to tell." *The New York Times Book Review* summed the book up as "[a] work of healing, of tranquil, even buoyant imagination." But alas, not so as Levi's suicide in 1987 was sadly to demonstrate. He could not shake the recurring nightmare focused so sharply in the poem that provides the epigraph for *Survival in Auschwitz*:

> You who live safe
> In your warm houses,
> You who find, returning in the evening,
> Hot food and friendly faces:
> Consider if this is a man
> Who works in the mud
> Who does not know peace
> Who fights for a scrap of bread
> Who dies because of a yes or a no.
> Consider if this is a woman,
> Without hair and without name
> With no more strength to remember,
> Her eyes empty and her womb cold
> Like a frog in winter.
> Meditate that this came about:
> I commend these words to you.
> Carve them in your hearts
> At home, in the street,
> Going to bed, rising;
> Repeat them to your children,
> Or may your house fall apart,

> May illness impede you,
> May your children turn their faces from you.
> (Levi 1961)

Sustained trauma can affect memory-loss, but all too often it can affect nightmarish memory-gain. And it is toward a consideration of memory-gain I want to turn through the impact not of the active trauma of torture, but through the impact of the passive trauma of solitary confinement. "Solitary confinement," one penologist said recently, "is the most diabolically severe punishment." And in passing we might remind ourselves that solitary confinement in sensory deprivation cells, twenty-two and a half hours per day, is the rule rather than the exception for the "hard cases" in the newest of our maximum security prisons.

One classic case of solitary confinement is amusing and instructive: in 1704 a Scottish seaman named Alexander Selkirk (1676-1723) quarreled with his captain and so seriously that he demanded to be and was put ashore on the nearest land which turned out to be the uninhabited island of Juan Fernandez in the Pacific Ocean some 350 miles west of Valpariso. No sensory deprivation there. The island was fertile and well-wooded; its highest peak was over 6000 feet, and from it Selkirk would have had a splendid panoramic view of the surrounding ocean from which he hoped for rescue. We wouldn't take much note of Selkirk but for the fact that his contemporary Daniel Defoe (c1660-1731) apparently did, transforming the unfortunate Scot into that intrepid survivor, Robinson Crusoe (1719). Selkirk spent four years and four months on the island, most of the time in a swivet of anxiety lest he be discovered by the Spanish and turned over to the tender mercy of the Inquisition and its torturers. As one commentator has remarked, "he had all the time in the world for bad dreams," and suffering them seems to have been his principal occupation, the beauty and serenity of his island environment lost on him.

Defoe presents his hero in a sharply contrasting light in that off-hand chatty prose of his, full of attention to the petty details of Crusoe's housekeeping and practical economy. I'll let James Joyce, a great admirer of Defoe's uninhibited middle-classness, sum Crusoe up for us; this is from a lecture Joyce gave in Italian to an Italian audience in 1912:

> The true symbol of the British Conquest is Robinson Crusoe, who, cast away on a desert island, in his pocket a knife and a pipe, becomes an architect, a carpenter, a knife-grinder, an astronomer, a baker, a shipwright, a potter, a saddler, a farmer, a

> tailor, an umbrella-maker, and a clergyman. He is the true prototype of the British colonist, as Friday (the trusty savage who arrives on an unlucky day) is the symbol of the subject races. The whole Anglo-Saxon spirit is in Crusoe: the manly independence; the unconscious cruelty; the persistence; the slow yet efficient intelligence; the sexual apathy; the practical, well-balanced religiousness; the calculating taciturnity. Whoever rereads this simple, moving book in the light of subsequent history cannot help but fall under its prophetic spell. (Joyce 1964: 24-25)

I find the contrast between the fictional Crusoe's practical equanimity and Selkirk's sustained nightmare instructive in the contrast it reveals between the way we imagine and relish (with Defoe's aid) the comforts of solitude and what the actual experience might otherwise be, implacably informed by things that go bump in the night of the imagination.

One other example: in 1949 the Hungarian-born writer and translator Edith Bone (1889-1975) who was an English citizen living in England was invited to Hungary to do a job of translation. Just as she was about to return home to England on the first of October, she was arrested, abducted really, by the Hungarian security police, a particularly paranoid bunch as totalitarian security police go. She was charged, on the basis of no apparent evidence, with being a "British secret service agent" (Bone 1957: 114). At first she was held in extremely unpleasant solitary confinement in order to wring a confession out of her, and after that effort failed, she was sentenced to "solitary confinement with total isolation." for an indefinite period (Bone 1957: 176). During the first few months she was harassed by every means short of active physical torture: sensory deprivation in dark and cold cells, sleep deprivation, interrogation at all hours, and four months of total darkness except at meal times. She was a determined and militant sixty-year old in her refusal to confess and in her resistance to what she regarded as absurdities in the prison routines and on the part of prison personnel, a resistance which she carried into a language strike, refusing to speak Hungarian or to understand her keepers. Eventually in 1952 she won the concession of her spectacles, writing materials and books from the prison library.

But during those first three years without print she kept herself sane by reciting poetry to herself, by translating it among the six modern languages in which she was fluent (English, Hungarian, Russian, German, French, Italian), and she took to composing poems of her own, what she called "doggerel" which she repeated to herself daily until the repetitions of her collected poems became too time consuming (Bone 1957: 132). She recalled a Tolstoi story in which a prisoner in solitary had occupied his mind by taking walks in cities he knew. She did the same and once walked herself from Budapest to London. She constructed an abacus out of pellets of bread and bits of straw. She took inventories of her knowledge and made vocabulary lists for the languages she knew. She started with English, in "strict alphabetical order" (Bone 1957: 131). It took three runs through the alphabet, and she reports, "I . . . found in the end that I had enumerated twenty-seven thousand three hundred and sixty-nine English words" (Bone 1957: 131). That abacus had apparently come in handy. And, she adds, "[t]hat satisfied me, and I went on to German, French and the rest" (Bone 1957: 131). She became, as she said, "an accomplished breadcrumb technologist" (Bone 1957: 132), creating a printing set with which she could lay out her poems on the table in her cell, and make Christmas and other seasonal decoration in addition to that abacus.

When after three years she was released into the world of writing and print (at the age of 62) her relentless enterprise continued unabated. She already knew Latin, so she taught herself Greek. She translated a Soviet book on interrogation and criminal psychology for the masters of her prison, and in so doing learned about the sorry state of Soviet psychiatry. She was forbidden all news of the outside world but got it anyway by taking the miss out of the misinformation that came her way in what she called "the long night of Russo-Hungarian inhumanity" (Bone 1957: 255). She emerged from that night by way of the window of opportunity provided by the all-too-brief Hungarian Revolution in 1956. She returned to England to bask "in the sunshine of civilization" when she was 67 after "[h]aving lived like a hedgehog, bristling in every direction" (Bone 1957: 212) for seven years, and her book *7 Years' Solitary* is a remarkable demonstration of her determination to keep her integrity and to use the power of memory to preserve her sanity of mind which was so severely under assault. She died in 1975 at the age of 86.

2. Art as a Way of Knowing

If memory is Mnemosyne, The Ur-Art, the mother of the Muses, then we are entitled to wonder whether She doesn't need (and have a right to expect) the support of the goddesses who are her children. This brings me to the role and function of the arts in our society. The arguments about the funding, or even the survival of the thirty-year-old National Endowment for the Arts (b. 1965) in the last several years have been instructive and, from my point of view, devastating in what they reveal about the poverty of mind of both sides in the debate. The opponents argue that the NEA is a waste of public money on things that are at best just frills (O.K. if you can afford them; if not, not). At worst, the opponents maintain that the frills the NEA supports are subversive if not pornographic and sacrilegious. The proponents, on the other hand, have tended to respond with their own version of cost-benefit analysis: *The New York Times* editorially reprimanded New York State's Republican delegates to Congress for opposing funds for the NEA because, the *Times* said, New York City's art institutions (many of which the NEA partially supports) are money-makers for the city; they attract tourists and tourist dollars. Let's keep government support for the arts to keep the local cash registers chiming.

And just over a week ago, *The Berkshire Eagle* [23 October 1995) weighed in with an editorial that quotes with approval the former high school principal Norman Najimy's outrage: "With school budget short falls, the arts are the first thing to go; they're considered a frill." The editorial then goes on to praise the Berkshire Cultural Alliance, The Central Berkshire Chamber of Commerce, and the Berkshire Visitors Bureau for their efforts in sponsoring and creating a resource booklet that lists the "museums, theaters, colleges, historical sites, and other cultural attractions" in Berkshire County. The booklet has been distributed to schools throughout the county to help teachers and their students enrich their experience of the arts and thus to compensate for the art-kill consequent on the budget short-falls. The pamphlets are also a good thing, the editorial says, because they will be "a boon to efforts to attract new business to the area." They "will be used as 'an economic development tool'" and will, it goes without saying, help to increase the tourist trade (*Berkshire Eagle* 1995: A6).

Great. Both editorials address their readers with the tacit assumption that most readers will not agree that the arts are a frill. But the cost-benefit approach implies something quite different: so what if the arts are a frill, they can help the City and the County sustain and improve their economies. In passing the *Eagle* editorial tips a hat to "those individuals that consider the business of living to involve more than the getting and spending of goods and services." I pause over that phrase "the business of living" in an editorial that drifts toward the argument that the arts are good for the living of business—as in Calvin Coolidge's immortal slogan: "The business of America is business." But another phrase catches me as well, "the getting and spending of goods and services"; and I wondered whether that editorial writer had intended an echo of the Wordsworth sonnet of 1807, the one that begins:

> The world is too much with us; late and soon,
> Getting and spending, we lay waste our powers;
> Little we see in Nature that is ours;
> We have given our hearts away, a sordid boon!
> (Wordsworth 1967: 536)

So whether that editorial writer was aware of it or not, that editorial stood briefly in the sign-stream of a 200 years old lament. And the following day, the *Eagle's* front page compounded the waste of our powers by announcing in the right-hand top news column that a State Senate committee had axed the sixteen million dollar fund for Phase Two of the Massachusetts Museum of Contemporary Art in North Adams. The left-hand column of that page announced approval of a twenty-six million dollar fund for a new prison in Berkshire County. The next day the cutoff of the museum's funds was called a misunderstanding and a mistake and restored. But the priorities of this society and its culture had spoken out loud and clear in that front page.

Time to start over, to start by asserting that nationally the underfunding and bureaucratization of the Endowments for the Arts and for the Humanities has been a scandal since their inception, and the present congressional determination to kill them compounds that scandal with outrage, or would if it were not a mercy-killing of something that's as wrong-footed as any other tourist attraction, wrong-footed because it assumes that "imagination and mind are politics" and, as Lionel Trilling said, "[politics] of a kind we will not like."

To start over again would be to redefine the National Endowments, to base them (and the school programs they should be promoting and supporting) on the principle that art is a fundamental way of knowing. Granted that art as a way of knowing has been much neglected in the last 200-plus years of western culture, neglected in favor of rational-analytic ways of knowing, plus, since the middle of the nineteenth century, statistical ways of knowing. The neglect has been so thorough that it is hard to know how to redress the balance between the now-dominant rational-analytical-statistical ways of knowing and art as a way of knowing.

I think of art-knowing as compounded of the intuitive, the affective, the mythopoeic aspects of human experience, with the proviso that those aspects of our experience, the abilities to intuit, to feel, to mythopoeticize can be schooled, trained, coordinated toward their own ways of getting things right, just as can our abilities to reason, to analyze, to count. What I am suggesting is that there can be good intuitions, good affections, good mythopoetics as well as unreliable or bad ones; just as there can be bad reasoning, bad analysis, bad statistics as well as good. And we ought to be about the business of training ourselves to make those evaluations, those distinctions in art-knowing just as we assume we must make those evaluations and distinctions in our rational-analytic-statistical ways of knowing.

"Science," Karl Krause (1874-1936) once said, "is spectrum analysis: art is photosynthesis." Science is involved in rational-analytical-statistical procedures, art in the immediacy of intuition. But as John Dewey objects: "Does not the reasoner have also to trust to his 'intuitions,' to what come upon him in his immediate sensuous and emotional experiences, even against objections that reflection [rational-analytic procedure] presents to him?" (Dewey 1934: 33). In other words, scientists frequently make important discoveries by intuitive leaps into the unknown from which intuitive vision they backtrack over rational-analytic paths of "proof" toward the starting point from which the original intuitive leap was made. (Incidentally, and parenthetically, this happens all the time in literary studies as the reader moves from "Oh, I see" to "Now, I have to reread the poem in order to demonstrate and prove my seeing.")

And so to John Dewey's definition of the function of art and its intuitive, affective, mythopoeic paths: "it . . . keeps alive the power to experience the common world in all its fullness" (Dewey 1934: 133). In Gerard Manley Hopkins's formulation: "it does this: keeps warm / Men's wit to things that are" (Hopkins 1986: "To What Serves Mortal Beauty?" 3-4). If we turn toward art as a way of knowing, we may find that it enriches the Ur-Art of Memory; we may also find better ways than the bean-counting studies of the so-called social sciences toward William James's ethical/aesthetic goal, an improved sense of "the inner significance of other lives," a significance which, as James said, "exceeds all our powers of sympathy and insight" (44). I pause here to note a coincidence, another divine pun: John Dewey delivered the lectures that were to become *Art as Experience* as the first William James Lecturer at Harvard in 1932. And William James's great *Principles of Psychology* (1890) moves in its leisurely two-volume-pace toward a final chapter the central emphasis of which is ethical-aesthetic ways of knowing. And a further pause: so much for Allan Bloom's *Closing of the American Mind.*

One of the saddest passages of autobiography I know occurs in Charles Darwin's unfinished and until 1958 unpublished autobiography. Darwin reports that "[u]p to the age of thirty, or beyond" he took great pleasure in poetry, "pictures," and particularly in music, but that now, late in life, "Music generally sets me thinking too energetically on what I have been at work on" (Darwin 1959: 138). He attributes what he calls "this curious and lamentable loss of the higher aesthetic tastes" to what he perceived as a mechanization of his mind: "My mind seems to have become a kind of machine for grinding general laws out of large collections of facts"; and he regrets not only the "loss of happiness" but also the possibility of injury "to the intellect, and more probably to the moral character by enfeebling the emotional part of our nature" (Darwin 1959: 139).

In the last two or three decades the cognitive sciences have turned away from a brief period of interest in art as a way of knowing toward a fascination with the algorithmic procedures of artificial intelligence, toward computers and information processing as models of the mind of this flesh. I wish they and the educators who look to them for guidance would turn back. I agree with the British educator James Britton (1908-1994) who says of one chapter in his book *Language and Learning*, this

> difficult and still unsatisfactory chapter, has two axes to grind: first to trace the development of cognitive processes, and then to attempt to right the balance by looking at the cognitive-cum-affective organization to be found in the arts, and applying the principles found here to more commonplace uses of language. Both these topics are of the utmost educational importance, but our ignorance regarding the second of them is positively dangerous. (Britton 1992: 6)

I agree that the chapter is difficult and unsatisfactory, but the urgency of the positive danger, the danger of our ignorance about art as a way of knowing, should outweigh the difficulties of Britton's style. We've been in this danger zone of ignorance for over two hundred years and should not be surprised that it is so hard to grope our way out. But the time has come.

Yet intuition can apparently still find ways to assert itself, to slither through the rational cracks. In Sarajevo a woman, herself traumatized by the siege and by the sniper-death of her husband, has opened a clinic which she calls a "drop-in center" for homeless and disoriented or traumatized children. She tries, as she puts it "to find a secret source of energy" in each child, and the means she uses in that search are play, music, and the arts ("All Things Considered" 27 October 1995). And I'm reminded once again of that tiny museum in the old ghetto in Prague which has on display drawings and watercolors made by children in the way-station concentration camp at Theresienstadt during the Holocaust: Downstairs, scenes of things wished for, family celebrations and dinners and the like, while upstairs, the grim present tense of their environment and in both, a feeling of a coming to grips with, an emotionally clear-eyed view of their appalling world.

And there are further questions: if we took art-knowing seriously as a way of nourishing memory, which as St. Augustine long ago remarked is "the stomach of the mind" (Augustine 1963: X.15, 224), would that effect change in the present sorry state of public education? Something like forty-five percent of the adults in Massachusetts cannot read or write with any degree of competence. That sorry statistic is attributed, in what is becoming an increasingly off-hand and reactionary way, to the failure of our school systems to hold their students to more rigorous standards. But I wonder if the failure isn't located in the pass-the-ketchup assumptions about language that dominate those "rigorous standards," that and a failure to nurture memory and its mind on the poetry it deserves?

And isn't it possible that the crazies and zealots among us who refuse to listen to reason might be calmed and persuaded by listening to a better art than craziness and zealotry? and that the poets and scientists among us could improve their poetry and their science by having their intuitive, affective, and mythopoetic abilities and skills trained and improved?

Where can the Muses go now? Toward an art that is the new that renews. Toward the Ur-Art; home to mother.

And so: I have done.

II

Time, Memory, and the Unconscious

Chapter Seven
The Imitation of Dream in Literature

Joyce described *Finnegans Wake* (1939) as an "imitation of the dream-state" (quoted in Ellmann 1982: 716n) and it is toward that description and toward the threshold of that work that these remarks tend.

Aristotle in the *Poetics* says that all literary compositions are rooted in "imitative processes" (Aristotle 1967: 15), but Aristotle's term *mimêsis* which we translate as "imitation" is a good deal more forceful than our term. *Mimêsis* implies not just a passive process of copying, replicating, representing but an active process of making and creating—a process of translating from one medium to another—a "carrying across" in the course of which the thing carried across undergoes an active sea-change.

For example, dialogue in fiction "imitates" conversation in "real life," but readable dialogue is rarely just a transcript of everyday conversation. In conversation there are so many dimensions (gesture, tone, histories, and contexts) that are not present on the printed page—conversely, the printed page enables us to retain, re-experience conversation as (mercifully?) we cannot in real life—consequently the literary re-presentation of conversation must be translated into style (a compound of the writer's own ear and mark and the literary deposits and traditions of the time). People in 1810 did not talk the way Jane Austen's characters talk (even though the thought that they did drove Mark Twain wild), nor did people in the Mississippi valley in the 1830s talk the way Huck Finn talks (in spite of Mark Twain's elaborate attention to phonetics and orthography); nor did lost generation types in the 1920s and 30s talk the way Hemingway characters talk—though after *In Our Time* (1925) and *The Sun Also Rises* (1926), it became fashionable to try.

The sea change that takes place in imitation, in the carrying over we call translation, subtracts several dimensions (including vagaries, incompletions, and repetitions) and adds others: form, concentration or compaction, and various levels of significance and resonance that add up to "interpretations of reality." Another way of suggesting the difference between conversation in "real life" and the imitation of conversation in fiction can be focused through Carlyle's dictum, "Narrative is linear; action is solid." In real life conversation is a multi-media event, an action which involves all the senses: not just hearing but seeing, smelling, all sorts of feelings, including the tactile and kinesthetic sense of speaking and listening, and the fabulous dimensions of memory. Linear narrative can never aspire to that sort of all-at-once density; it can however translate into densities of its own.

Another inadequacy in our terminologies bears on my topic, the imitation of dream: we frequently use the phrase "Stream of Consciousness" to describe a narrative method in the fiction of the last one hundred years or so (pointing to George Meredith, Henry James, Edouard Dujardin, Dorothy Richardson, James Joyce, and a host of others). But what the term neglects is that we are talking about *imitation* (re-presentation, translation) of stream of consciousness, not the stream itself. Consider the origin of the phrase in William James, *The Principles of Psychology* (1890):

> Consciousness, then, does not appear to itself chopped up in bits. Such words as 'chain' or 'train' do not describe it fitly as it presents itself in the first instance. It is nothing jointed; it flows. A 'river' or a 'stream' are the metaphors by which it is most naturally described. *In talking of it hereafter, let us call it the stream of thought, of consciousness, or of subjective life*. (1:239)

In literary works we do not meet the stream of consciousness itself but some imitation, representation of that stream—frequently in prose passages which are in themselves remarkably fragmented and discontinuous—as in the opening page and a half of Joyce's *A Portrait of the Artist as a Young Man* (1916) where a sequence of nine or ten fragments provides a comprehensive metaphor for the small child's stream of consciousness—verbally it is all chopped up, but the narrative imitates (and we read it as) flow. William James was aware of this sort of all-chopped-up discontinuity in consciousness itself; his point is "that within each personal consciousness thought feels continuous" and if "there is a time-gap the consciousness after it feels as if it belonged together with the consciousness before it" (James 1890: 1:237). And as readers of stream-of-consciousness fictions we knit-over time-gaps (Jaynes 1976: 25) in the narration as we knit-over time-gaps in our own consciousnesses. Elsewhere James suggests the other, all-but-inimitable condition: that consciousness is not a single stream but many-streamed, a delta with many channels, backwaters, and eddies.

And so to the imitation of dream: in our everyday accounts of our dreams we usually rely on linear narration. After the fact of a dream we translate the dream into a story. But since dreams are (or at least the ones we remember are) more often than not visual—the solidity of the visual and kaleidoscopic imagery of dream (what Lessing called the *nebeneinander,* one beside another) must give way to the linearity of what Lessing called the *nacheinander,* one after another, in the telling (Lessing 1962: 77). The density of poetic language can, of course, achieve its own sort of *nebeneinander,* but it is delightfully not the same as the *nebeneinander* of visual imagery. I've already suggested that the dimensions of that not-the-same are the various levels of significance that add up to our interpretations of "reality." And there are in our dream-traditions fabulously rich deposits of interpretation.

In the Bible, dreams are either vanity and false prophecy (sent by a devil or demon to tempt and corrupt) or God-given prophecies of the future—as Joseph in Genesis 40:8, "Do not interpretations [of dream] belong to God?" and when, in Genesis 41 he interprets the Pharaoh's dream of the seven fat and the seven lean kine, he says "God hath shewed Pharaoh" (41:25).

In the second chapter of Daniel there is a notable catch-22 example of the divine nature of dream-interpretation. Nebuchadnezzar, king of Babylon, has had a dream but cannot remember it. He summons his magicians and demands (on pain of death) that they remember it for him and then interpret it! The secular magicians can't hack it, but the secret of the king's dream is "revealed unto Daniel in a night vision" (2:19). The dream Nebuchadnezzar had forgotten was the famous one of the great and terrible image "whose brightness was excellent" (2:3 1) but whose feet were "part of iron and part of clay" (2:33) and therefore vulnerable (as we remember not in dream but in the feet-of-clay cliché) to destruction as Babylon was vulnerable.

A similar ambivalence about the prophetic validity of dreams is evident in Greco-Roman tradition as in the *Aeneid,* when the ghost of Aeneas's father Anchises says to his son Aeneas in what we read as "true prophecy":

> There are two gates of Sleep, one said to be
> Of horn, whereby the true shades pass with ease,
> The other all white ivory agleam
> Without a flaw, and yet false dreams are sent

> Through this one by the ghosts to the upper world.
> (Virgil 1983: IV:893-99)

During the first millennium B.C. an ancillary tradition emerged through the compilation and circulation of "dreambooks," the most notable among them, the Assyrian ZIQIQU which spawned countless imitations. They functioned (and in debased forms continue to function) as code books that promise to decipher dreams as though they were universally coded messages.

And those two contrary traditions (true dreams/false dreams) continue in various guises right down to our own day: that dreams are somehow true and prophetic, or that they conceal and delude. That they can be read straight; that they go by contraries; that they should be ignored—not all that hard to do since we so readily forget our dreams unless we train ourselves to remember; and some modern dream researchers and theorists are of the opinion that we are meant to forget, that forgetting is part of the mechanism of healthy dreaming.

During the nineteenth century, conventional scientific wisdom held that dreams prevented the interruption of sleep, that they were aroused by and connected to sensory stimuli: "light dust" on the retina at dawn would spark the visual images of dream, a skin irritation would immediately translate into a dream of insects, and so forth. Dreams were also assumed to be "predominantly visual."

Freud's *The Interpretation of Dream* first published in 1900 has been a landmark document for our century. Freud agreed with his contemporaries that the function of dreaming was to preserve sleep from interruption. But he argued that dreams were triggered not by somatic events but by potentially disturbing psychic events which have their origin in the unconscious. What he called the "manifest dream" (the dream we remember to one degree or another) disguises, censors, chaperons the "latent content," and Freud asserted that when the latent content was revealed, "When the work of interpretation has been completed the dream can be recognized as wish-fulfillment" (207). The wishes encoded in dreams are not just wishes in the present tense of one's life, but wishes from the whole of the individual's psychic history, recapitulating the personal history of wish from its origins in the wishes of repressed infantile sexuality. Freud's operating assumption was that dreams are not only wish-fulfillment but also that all dreams are "full of significance," significance that can be "assigned to a specific place in the psychic activities of the waking state" (183). The psychoanalyst, with the help of the patient, dissects the dream so that each of the elements of the dream can be "assigned to a specific place." That cannot, Freud initially argued, be done with the "literature of dream life" because the patient is not present and therefore the analyst is "deprived of the analysis" (194). Freud did not stick to this view, and his early reservations have not, of course, prevented psychoanalytic criticism (and sometimes composition) of our literatures, but I mention those reservations as part of a larger cautionary tale.

Contemporary dream researchers (and their theories or rather hypotheses) have not challenged Freud's basic assumption about the psychic importance of dreams, but it is far from clear to them that the function of dreaming is to preserve sleep or that dreams are psychic "safety valves." Dreams can of course be analyzed for their latent content and the relation of that content to waking consciousness as before because that is primarily a function of the waking consciousness as it develops that latency; but dreams seem to be triggered not by some psychic disturbance originating in the unconscious but by some sort of physiological clock in the brain.

Largely on the basis of electro-encephalographic readings (EEG), contemporary research outlines four levels of sleep, each with its characteristic dream states: First, Rapid Eye Movement (REM) sleep. The dreamer has an EEG similar to that of the waking state; so much so that some researchers call this "the other state of consciousness" (Winson 1985: 35). The primary difference between waking consciousness and REM-sleep is that during REM-sleep the neocortex is isolated from the external environment and from muscular control of the body.

In the second and third stages of sleep the EEG shows a less active brain; the content of the dreams reported by subjects awakened during those stages is much less visual. They characterize their dreams as "thinking" dreams, dreams of reading, and what one researcher calls "wasteland dreams." During the fourth stage of sleep the EEG is like that of a person in a coma—and still there are reports of "wasteland dreams." Most people pass through five to seven cycles of all four stages each night, spending about twenty-five per cent of the night in REM sleep, twelve per cent in coma sleep. You would imagine that it would be difficult to awaken people from coma-sleep, and it is, but paradoxically it is easier to awaken people from stages two and three than from REM- and coma-sleep.

The question: what is the function of dream is still a vexed and vexing one. But we do know that sleep-deprivation (dream-deprivation) is one of the most insidious forms of torture, and we can thus conclude that dreaming is fundamentally necessary to psychic health. Observers engaged in dream research report that when the subject is wakened, "the dream dissipates rapidly in the time between sleep and full awakening" (Winson 1985: 51). This has led to a number of hypotheses; that the recall is not necessary to the primary function of dreaming, that too much recall might even interfere because the primary function of dreams may be "to integrate experience over time" (Winston 1985: 205); that what we are witnessing—or rather eavesdropping on—in our dreams is the sorting, winnowing, integrative process which is the making of memory; and that, in order to accomplish this integration, consolidation, catharsis, and clarification dreams super-concentrate experience and response in elaborate symbols that are expressive rather than repressive. Repression is the contribution of the volunteer-subject in dream research, the patient under analysis, and in many ways the observer or interpreter.

Freud and a host of others have assumed that all dreams are primarily visual. In an all-night sense that is apparently not the case, but the last installment in most people's every night sleep-time is an REM stage; so the highly visual dreams of that last REM installment are the dreams that, upon awakening, most of us most frequently recall. As in all research, the presence of an observer or an analyst interferes with and distorts the thing being observed. The psychoanalyst who "rewards" a patient's dreams with special attention when the latent content is demonstrably sexual may in effect be teaching his patient what to dream or what to remember or what to report; and the dream-researcher, pad or lab telephone in hand, may teach the dreamer to variously censor, to dull the jazz of his dreams—not consciously in the reporting or unconsciously in the dream itself but in the mediating presence of memory and recall.

In all of this, there is something suspicious. Throughout history human beings have asked (as we continue to ask): what is the function of our dreams, what do they mean? with the proviso: and therefore, how can we use that other state of consciousness to improve our waking lives? But why do we not, anything like as consistently, ask those questions of our waking lives? That inconsistency leads me to wonder not only about the relation between our waking and sleeping lives, it also leads me back to questions about consciousness itself (and its well-publicized stream). So much of what is thinkably available to the daylight of consciousness passes without our notice, so much remains on the periphery where we do not even bother to glance at it. And yet what we discover in dream (recover from dream) is a vocabulary of notice that goes well beyond our waking capacity to envision, glance, and recall. The quietest of our days conceals a riot of stimulus and response: unconscious response, pre-conscious response, conscious response. Of the fabulous range and quantity of stimulus much is filtered out, shunted aside, repressed in a process of editing which, considering the buzz and din total awareness would threaten, is essentially benign. But we rarely pause to examine and question that editorial process, to wonder whether we are filtering out the gold and merely retaining the dross, whether we are shunting aside what would nourish, whether what we repress when we are cruising on automatic pilot is what really merits our attention.

I have already suggested that the "imitation of consciousness" in what we call "stream of consciousness" implies a linear simplification of what might better be called the "delta of consciousness," and I would argue that imitation of the delta of consciousness is most successfully realized in James Joyce's *Ulysses* (1922) and that the imitation of dream is the imitation of the delta of unconsciousness or semi-consciousness at the compound interest rate of *Finnegans Wake:* and that brings me by one route to the threshold and conclusion of my subject. An alternate route toward the same threshold might begin with literary models: fairy tales and Medieval Dream Visions. Fairy tales, which we associate with folk tradition and childhood, have a quality-texture that informs our dreams in childhood and which we readily relate to the world of adult dreaming (and vice-versa).

2. Dream Visions

Medieval Dream Visions are in the mainstream of our formal literary heritage. They ostensibly take place in a dream state, but they are intensely allegorical in content. If they can be said to be imitative of dream at all, their imitativeness is rooted in the assumption that dreams present truth in disguise, that dreams are by their very nature allegorical. But in practice, Dream Visions are much more imitative of each other and of their classical and Biblical models than they are of the experience of dreaming. The dream state, once announced, functions as a frame, cutting the world depicted off from the continuities of "real life," urging it as the other-than-life of allegory, whether an allegory of love as in that most influential of Medieval Dream Visions *Le Roman de la rose*, or an allegory of religious experience as in Dante's *La Divina Commedia* and Langland's *Vision of Piers Plowman*.

In most Medieval Dream Visions what the reader has always known is being urged not just for the sake of reassurance or agreement but to point out that the tongue's ready assent to the *truth* of Christian doctrine or the *truth* of chivalric love is not necessarily the heart's living participation in those truths. In effect the allegorical content of Medieval Dream Visions is not as straightforward and unambiguous in "message" as it has frequently been described by those readers who are impatient with allegory that seems too obvious. At their best, Medieval allegories are demanding moral and intellectual exercises. And consciously so. In the letter dedicating the *Paradiso* to his patron Can Grande della Scala Dante quotes from Psalms 114:1:

> When Israel went out of Egypt, the house of Jacob from a people of strange language,
> Judah was his sanctuary, and Israel his dominion . . .

and then Dante details the four levels of allegory in the light of which that passage is to be read:

> Now if we look at the letter alone [the literal sense], what is signified to us is the departure of the sons of Israel from Egypt during the time of Moses; if at the allegory . . . our redemption through Christ; if the moral, . . . the conversion of the soul from the sorrow and misery of sin to the state of grace; if . . . the anagogical, . . . the departure of the sanctified soul from bondage to the corruption of this world into the freedom of eternal glory. (Dante 1973: 99)

To trace the four levels of allegory in the *Divine Comedy* places severe intellectual demands on the reader, but the aesthetic excitement of the poem is inextricably bound up with the intricate choreography of its allegory and its ideas. A capacity for excitement about the play of ideas and the play of wit is indispensable to an appreciation of Medieval Dream Visions, as is a willingness to contemplate the moral disparity between what one knows to be Truth with a capital T and the twilight of half-truths in which we pass most of our nights and days.

After the Middle Ages the Dream Vision fades toward neglect though it is vigorously alive and well in Bunyan's *Pilgrim's Progress* (1675) which is presented "under the similitude of a dream." Ideas (and the preaching of them) are still in the fore-front, but powerfully rivaled in our interest by Bunyan's narrative energy and inventiveness. *Pilgrim's Progress* is a rattling good story which survived into my childhood not for the precision of its doctrine but for the excitement of its geography, its cast of characters, and its story. Much the same thing could be said about those splendid nineteenth-century Dream Visions, *Alice in Wonderland* (1865) and *Through the Looking Glass* (1871). Elements of nonsense, satire, and parody inform their compelling narrative energy, but they are not, in any sense, sustained by the allegorical density of their ideas.

Hawthorne had some fun with Dream Vision in a mode that might be called "anti-allegory" because he undercuts his audience's preference for the easy moral lessons of cut-and-dried allegory in favor of an allegory that explores historical and psychological ambiguities. In "Young Goodman Brown" (1835), for example, Brown has a vision of a witch's sabbath in which he (and his wife Faith) almost participate. When the vision evaporates the unanswerable question remains was it real or was it dream? In any case the psychological impact of that night-experience sours his life with the prideful assumption that he alone has not been tainted with sin. The adventure of one night's brush with evil becomes an ambiguous allegory of a gloomy coming of age, and an ambiguous historical allegory which outlines the gloomy presence of the Calvinist doctrine of the total depravity of man and the unsympathetic self-righteousness which that doctrine can breed in those who suppose themselves righteous.

Shakespeare's *A Mid-summer Night's Dream* (1595-96?) runs comic variations on Medieval Dream Visions. The frame is the court of Athens, a sort of Elizabethan Great House, presided over by Theseus and Hippolyta (his wife to be) who represent the practical and responsible intellect of the aristocracy. The forest outside Athens is the world of dream, housing variously three groups: four mixed-up young lovers; Oberon and Titania (king and queen of the faeries whose quarrels locate them on the verge of divorce); and the "rude mechanicals" (Bottom and his fellow workmen who are preparing an entertainment for the wedding at court). The dreams are not allegorical but the dream-teams do act out overlapping and contrasting comedies of misrule which in the rational light of day Theseus can dismiss with the much quoted lines: "The lunatic, the lover, and the poet / Are of imagination all compact" (Shakespeare *A Mid-summer Night's Dream*: V:i:7-8); Hippolyta's yes-but answer is less frequently quoted, but hers is the sensible observation that the imaginings of that triad of dreamers do make a distinct difference in their lives and quite possibly in the lives of others who are awake. She tells us, in other words, where the play lives and what it lives on—dream as a sustained metaphor for the human imagination in action.

Coleridge's "Kubla Khan" is a highly concentrated metaphor for that same action. It is composed of three parts: First, the headnote (ostensibly added in 1816) that explains that the poem was composed in 1797 while he was in an opium dream; that he began to write it down when he awoke, but that he was interrupted by a notorious spoil-sport since known as a "person on business from Porlock," and that the long balance of the poem was consequently lost (Coleridge 1985: 102). The second part of the work, lines 1-36 of the poem proper, is an unmediated vision which begins: "In Xanadu did Kubla Khan / A stately pleasure dome decree" and ends, "It was a miracle of rare device, / A sunny pleasure dome with caves of ice!" The third part is a mediated vision: A damsel with a dulcimer / In a vision once I saw" (lines 37-end, Coleridge 1985: 102-104).

In the *Biographia Literaria* (1817) Coleridge says that he and Wordsworth agreed in 1797 that his role in the composition of *Lyrical Ballads* (1798) would be to experiment with demonism and the supernatural as vocabularies of metaphor for psychological states he described as "deeper even than our consciousness" (Coleridge 1951: 433). And the presences in the poem of the "demon-lover" (line 16) and the demonically possessed poet suggest that the focus of the poem is that deeper-than-consciousness thing, the poetic imagination.

There are, then, two ways to take the headnote: literally, in which case we believe in the man from Porlock on business, and believe Coleridge's assertion that there was much, much more composed, and we accept his dismissal of the poem as "a psychological curiosity." Or we can take the headnote figuratively, and regard it as an integral part of the poem, suggesting an order of psychological experience that can be imagined but can never truly be, the dream-state in which the poem was composed: "if that indeed can be called composition in which all the images rose up before him as *things*, with a parallel production of the correspondent expressions, without any sensation or consciousness of effort" (Coleridge 1985: 102). What Coleridge is describing is Orphic utterance in which words and images and things are one and the same. When Orpheus sang that the rocks and trees danced, the rocks and trees danced. But while that trance-state can be imagined, it can never be. Human "vision" is evanescent, and Xanadu as dream vision is a vision within a vision within a vision. The vision of the true imagination would be a-referential—no temporal or spatial discontinuity would remain between dream and reality— and that is a waking impossibility.

If we turn to the traditions of the novel, we find that the style in which dreams are presented tends to be consistent with the style of the novel in which they are embedded—literally and figuratively the dreams remain "in character"—close to day-dream and at times interchangeable with it as in Ford Madox Ford's *The Good Soldier* (1915) when the narrator has a vision of the dead wife who has so devastatingly betrayed him and his friends. She is isolated, "intolerably solitary" on an "immense plain" which is "the hand of God" (Ford 1951: 70). The language and imagery are consistent with the language and imagery the narrator employs to describe past and present scenes and his conscious responses to them. In effect, the style established as appropriate to the narrating consciousness of the novel is interchangeably appropriate for day scenes and night dreams.

Dreams function in various ways in the traditions of the novel. Here are some examples, though hardly a catalogue: In the Ford Madox Ford novel the dream is literally and figuratively a concentrated metaphor for larger phases of the narrator's bewildered and groping consciousness. In Dickens' novel *Hard Times* (1854), Stephen Blackpool, a mill hand, is ground down by an alcoholic wife, and, torn because his life is at the same time illuminated by his forbidden love for Rachel, a saintly fellow mill hand. He has a dream vision of his death by hanging (from his loom, appropriately enough) and at novel's end he meets his death by falling down an abandoned mine-shaft, the Old Hell-Pit, another symbol of the heartlessness of the world created by the industrial revolution. His dream and his death combine as concentrated metaphors for what the industrial revolution has done to impose barren lives on the working class, lives deprived of any sustained comfort from or nourishment by human kindness.

In Emily Bronte's *Wuthering Heights* (1848) the narrator Lockwood is storm-bound at the Heights and is accommodated in a closet bed. He has a three-phase dream: (1) He is mesmerized by the writing on the window sill of the closet bed, and the words play drifting-off-to-sleep games in his head. (2) He dreams of a sermon of four hundred and ninety parts in a country chapel. The tone images the grim tone of the kitchen he has just left and the sermon ends (appropriately) in a riot of misrule, aimed at him but spreading to all versus all, "Every man's hand was against his neighbor"—an accurate metaphor for the moral climate at Wuthering Heights where he is marooned. (3) A branch is scratching against the house. Lockwood dreams that someone is scratching at the window and puts his hand out only to have it grasped by an icy-cold child's hand. The heroine Catherine Linton has come back from the grave, home from twenty years lost on the moor. Lockwood cries out in his sleep. Heathcliff (the novel's demonic hero) storms into the room and responds to the dream as though it were true because in the demonic world he shares with Catherine, it is true! His irrational response contrasts with Lockwood's limp-wristed skepticism to underscore the demonic status of the novel's hero and heroine, a status Lockwood will never be able to imagine or to understand.

In Melville's *Moby Dick* (1851) there are three notable dreams. In the first, Ishmael the narrator is pinned down by Queequeg's arm which has been thrown over him in sleep. He eventually perceives his predicament as "comical" and recalls a nightmare he had had as a child when "a supernatural hand seemed placed" in his and he had experienced a spell of terror which he could have broken by stirring "a single inch," except that he could not stir (820-21). Melville uses this scene to image one of Ishmael's saving graces, his ability to accept and contemplate the conjunction of comedy and terror.

In chapter thirty-one of the novel there is a comic dream presided over by Shakespeare's invention "Queen Mab," the fairy who cheats human beings with dreams. The dream is told by the second mate Stubb who regards all of life as a vast practical joke at the expense of man, a joke to be dissolved in laughter, and his dream is just that, an "extravaganza," full of literary allusions, slapstick, and a comic argument with a fatal consequence: it enables Stubb to accommodate himself to Ahab's demonic tyranny—and thus, in the end, to accommodate himself to the destruction of the ship and the crew and himself (931-32).

Later in the novel another dream focuses kindred ideas in a far more serious way. In chapter ninety-six "The Try-Works," Ishmael at the tiller gazes into the midnight fire of the try-works until he falls asleep and turns back-for-front (his back to the compass and to the bow of the ship). His confusion almost capsizes the ship. Righted, he interprets his dream: "Look not too long in the face of the fire"; to do so means permanent disorientation (1247) We in turn read Ishmael's insight as a commentary on the novel's demonic protagonist Ahab who has been permanently "inverted" by too sustained a contemplation of the Inferno in himself and who destroys himself and his ship and his crew by projecting that inferno onto and externalizing it as the fiery quest for Moby Dick.

As these few examples suggest, dream scenes can be extraordinarily effective novelistically as concentrated metaphors for larger phases of the work, its atmospherics, the mental and emotional states of its characters, and the novel's play of ideas; but the dreams remain scenes within the narrative fabric of the novel as a whole, not appreciably different in style or function from all sorts of other scenes: descriptions of settings, of actions, of characters, narrative presentation of dramatic encounters, passages of stream-of-consciousness. In other words, dream scenes in novels do not imitate the stuff of dreams themselves as much as they imitate the vernacular habit of dream recall—a habit as artificial as our habit of telling stories of things we remember as chronological sequence when memory itself retrieves the past not as chronological order but as temporal hodge-podge. And narrative, as we read, almost always urges upon us the implication that it is being authored—as when we story our dreams, we assume the role of author of our dreams. As dreamers we witness and act in our dreams as spectators rather than as authors. Even when we are victims, we are somehow members of an audience of one rather than the active agents of our victimization.

In this respect the imitation of dream in the drama has the advantage that the dream (as the play itself) is presented directly, and gives the illusion of being authorless. There have been notable experiments with the imitation of hallucination and dream in the drama. Gerhardt Hauptmann's 1892 play *The Assumption of Hannele* carries the subtitle *A Dream Poem*. Hannele, an abused and neglected child, is on her deathbed. The apparitions she experiences in her delirium are introduced as characters on stage. The effect: two contrasting orders or classes of character—the "real" characters in the present time of the action and another group, the shades or hallucinations in their extra-present tense, dramatically just as present on stage as the real characters but disturbingly "different." Strindberg explored similar techniques in several plays, most notably *The Dream Play* (1902) and *The Ghost Sonata* (1907), in which he achieves a similarly disturbing effect of double vision.

These techniques Joyce borrowed and adapted to the novel in the Circe episode of *Ulysses*. Typographically, the episode is presented as a play with two levels of action: everyday reality on the one hand, and, on the other, extended hallucinations which we associate with one or another of the "real" characters but of which the characters themselves are only fragmentarily, dimly aware. And the two levels interconfuse, blend into each other so that we often do not know whether to place a moment in reality or in hallucination. Joyce once defined his "object" (what he wanted to imitate) as "sensation, even to the point of hallucination" (Power 1974: 67), and he said he had come close to that object in the Circe episode. The technical difficulty: true hallucinations, for the person experiencing them, are indistinguishable from reality; and the reader must be able to perceive the character's experience of that "indistinguishable" and at the same time be able to make the distinction. Reading Circe, it is as though we were witnessing the action in two different literary forms at once—as though we were simultaneously present at a play and reading a novel. In the theatre the clock that times the dialogue on stage more or less matches the time on our wrist watches; but the novelist is free to go on for pages to develop the resonance of a moment's action, an instant of hallucination. What Joyce manages is to combine the advantages of both drama and novel in the imitation of dream/ hallucination.

One more medium for raising questions about the imitation of dream: film. Jean Cocteau called film "the great machine of dreams"; George Orwell once suggested that films show us our dreams; and there is the ever-present cliché that Hollywood is, or at least was, the dream factory. But this widely perceived dream/film linkage has its ambiguities and excites a number of questions: Do films show us our dreams? that is, do they somehow do the dreaming for us and leave us only the effort of bearing witness—as Hollywood once packaged daydreams of glamour and success for us, dreams of what Ogden Nash called "a life of nonchalance and insouciance"? And does the concentrated visual imagery of film resonate with the intense visual condensation that most of us recall from our dreams? And how much does the imagery of film infiltrate our dreams?

Is Cocteau suggesting that film manufactures dreams for us, or is he suggesting that film is the ideal medium for the imitation of dream? And why do films imitative of dreams (including Cocteau's) present dreams as narrative and/or as complex choreographies of surreality and symbol? Dreams in experience are rarely inherently narrative until we translate them into narrative in the process of recalling and telling them. And dreams become symbolic not in and of themselves but in our experience of interpreting them: whether in the classical and Biblical traditions of reading them as omens, or in the Freudian and post-Freudian traditions of relating them to "the psychic activities of the waking state" (Freud 1939: 183). In short: why has film, for all its asserted promise, achieved so little in the imitation of dream?

The dreams most of us recall are predominantly visual, but there are those rare few among us who report to dream researchers that their dreams are predominantly verbal, more like listening to the radio than being at the movies. And if my impression is right, that historically the world of the ear was once in the ascendant, as for example in Shakespeare's time, did many more of those ear-minded Elizabethans have listening dreams instead of looking dreams?

And so we come by this alternate route to the threshold and conclusion of my subject: the assertion that the finest "imitation of the dream-state" in our heritage is Joyce's *Finnegans Wake*. Joyce's intricacies are, of course, verbal, not visual, but he did say in conversation: "Indeed you can compare much of my work to the intricate illuminations [in *The Book of Kells*]" (Power 1949: 67). And when *The Book of Kells* makes its appearance in *Finnegans Wake* (Joyce 1939: 122.23), the implication is that the *Tunc* page, one of its most famous illuminated pages, has been inspired by a passage in *Finnegans Wake* (which, because it's a dream, had its being in the dawn of prehistory and thus predates the eighth century *Book of Kells*).

Speculation that Joyce dreamed in verbal rather than visual imagery is, in the absence of firm evidence, idle; but Ellmann reports that Joyce was fascinated by other people's accounts of reading and vocables in dreams and by dream language. A year before I encountered *Finnegans Wake*, I had and recorded a couple of language dreams that were composed of strange amalgam words, English/French/German/Italian/pidgin-Arabic. When I began to read *Finnegans Wake*, I felt that that sort of amalgam language, remembered and elaborated into a principle of language-making, could be the basis of a whole lexicon for the imitation of dream. The extraordinary verbal tapestry of *Finnegans Wake* is, I suspect, woven that way from the seventy "foreign" languages that make up the weft which is worked into the hard warp of English that underpins the tapestry with its vast vocabularies (including archaisms, dialects, and slangs).

In 1919 Lord Rutherford split the atom; in *Finnegans Wake* Joyce undertook what he called, in *Wake* language *"The abnihilisation of the etym"* (353.22)—e-t-y-m, to the roots of language in etymology and beyond. Many commentators (extracting one pun from "abnihilisation") maintain that Joyce's night-language "annihilates" the word, but the pun also suggests that this is a language created "ab nihil," creation from nothing (as presumably only God can create). Joyce said to Jacques Mercanton, "I made it [*Ulysses*] out of next to nothing. *Work in Progress* [FW] I am making out of nothing" (Ellmann 543n). The result is a language appropriate to the imitation of dream—not imitation through after-the-fact narration or symbolic interpretation, but an imitation of what the experience of dream would be like if we could witness our dreams as they unfold in the present tense of sleep instead of in the past tense of awakened recollection. And, if that is where we are in the *Wake* (in the bleachers at the event itself) then the question that vexes so many Joyce scholars "Who is the dreamer?" disappears because only very rarely is there a "who" who is dreaming in our dreams. As dreamers, we witness, but only when we are awake and telling the story of a dream do we author.

One final observation: in spite of what many cultural critics see (and lament) as the decline of the verbal, the ascendancy of the visual in our eye-minded culture, it may still be possible that the ear-mind lurks just beyond the range of our peripheral vision, ready to slip through the gate of horn "whereby the true shades pass with ease," ready to redress the imbalance through that densest of verbal tapestries *Finnegans Wake*. It may still be possible that through the language of dream the ear will be reborn to and realize again a Shakespearean dominion. For that to happen, the *Wake* would have to colonize us, infiltrate, transform, refocus our sense of language and dream. Impossible as that colonization seems when we consider the monumental difficulty of the *Wake*, we should remind ourselves that seventy years ago T. S. Eliot's *The Waste Land* (1922) and Joyce's *Ulysses* seemed equally difficult, and yet they have colonized us quite effectively, thank you. They have become nature to us—as I'm sure *Finnegans Wake* will when we stop treating it as the world's biggest double-crostic and stop trying to translate it into a conventional "novel" with cast of characters, scenes, plot, and so forth, and treat it as the immediate dream-witness, the imitation of dream and the extraordinarily courageous accomplishment which it is. As Jorge Luis Borges said of Joyce:

> He comprehended that the effort to mould the incoherent and vertiginous matter dreams are made of was the most arduous task a man could undertake, though he might penetrate all the enigmas of the upper and lower order: much more arduous than weaving a rope of sand or coining the faceless wind. (Borges 1964: 74)

Chapter Eight

The Chip on His Shoulder: One for the Joyce Centennial 2 May 1982

On the second of May 1934, Joyce wrote to his friend Frank Budgen to commiserate with him on the lagging sales of Budgen's book, *James Joyce and the Making of Ulysses*. Joyce summed up: "I have a grocer's assistant's mind" (Joyce 1966: 3: 304). I like to take that sentence out of the context of book sales and read it as an allusion to the thousands upon thousands of scribbled bits which Joyce accumulated in his ever alert inventory of slips of the tongue and resonant puns, the inventory which was to be layered into the linguistic and mythopoeic exuberance of *Finnegans Wake*, because exuberance is the word for Joyce, and I'm delighted that emphasis on his exuberance seems to be in the ascendant.

Now to my title: "The Chip on His Shoulder": Several years ago my wife and I were vacationing on Inishboffin (the Island of the White Cow) off the west coast of Ireland. An Irish nurse from the Constance Markiewitz Hospital in Sligo, the Connacht seaport Yeats so loved in his childhood, was stormbound on the island with two friends. When the subject of Yeats came up, the nurse from Sligo said: "I don't understand why you Americans are so interested in Yeats. After all, he's not one of us!" In a stroke she had removed the Anglo-Irish from the Irish literary heritage: gone were Swift, Sheridan, and Burke . . . . Wilde, Yeats, Shaw, and Synge, and a good many Irish patriot-heroes along with them, from Wolfe Tone and the United Irishmen of the Rebellion of 1798 to Charles Stewart Parnell, the giant of Irish politics in the late nineteenth century. Perhaps Joyce could be salvaged. I tried, but he too met rebuff: "*He* had a chip on *his* shoulder!"

I was later to realize that in the west of Ireland all Dubliners are suspect as people with chips on their shoulders. But Joyce stands taller, and his chip weighs more, not only in the west of Ireland but also in Dublin where many are still inclined to regard him as a cold man with, in his phrase, a "coldsteel pen" (Joyce 1986: 1.153) in his hand, and they seem to fear, as he hoped they would, "the lancet of his art" (1.152).[1] Dublin anecdotal history still clings to (or at least improvises for tourists) stories of Joyce's ominous anti-social behavior. He used to sit in pubs, they will tell you, and eavesdrop on other people's conversations. He even took notes, and, when he escaped abroad, he had with him (depending on the informant) anything from a suitcase to a steamer-trunk full of those notes, a life-time supply of the materials he would use to betray the confidence and distort the lives of his countrymen.

After all, he did dismiss his countrymen as "the rabblement of the most belated race in Europe," and with slashing irony he depicted Dubliners and their institutions as paralyzed by simony. The words simony and paralysis are given pride of place in the first story of *Dubliners*. Joyce was fond of using the word simony figuratively, as it was used in the Medieval Church, to mean the prostitution of any spiritual value, of any of the seven gifts of the Holy Ghost (wisdom, understanding, counsel, fortitude, knowledge, piety, fear of the Lord). The Church, so revered by Dubliners, Joyce portrayed as, at its worst, fear-mongering and psychologically brutal and, at the same time, infected with the sentimentality and soft-sexuality of Mariolatry. At its best, he portrayed the Church (in the person of Father John Conmee in *Ulysses*) as wandering in a slack-minded, quasi-liberal meliorism. Once, in *A Portrait of the Artist as a Young Man* (chapter 3:C^2) a Capuchin priest, surprised and disturbed by the sixteen-year old Stephen Dedalus's confession of sexual precocity, responds compassionately to the terrified young man. That gentle response should have come as something of a surprise to Joyce's Dublin contemporaries because, at the beginning of this century, the Capuchins were well known in Dublin as tough and rigorous in the Confessional. But that moment of compassionate kindness in *A Portrait* is overshadowed by the hellfire stridency of the four sermons that precede Stephen's confession and by an outstanding example of the pride and arrogance of the priesthood just a few pages later (chapter 4:B). And Joyce's attitude toward Irish politics was equally scathing. His brother Stanislaus said that Joyce "reproached his contemporaries for being extremely heterodox in thought, politically orthodox in action, and revolutionary only at the beer table" (Stanislaus Joyce qtd. in Joyce 1975: 15).

Understandably many Dubliners have had a go at that chip on his shoulder, challenging Joyce's right to wear it as jauntily as he did and does. Until a few years ago the need to challenge was not as urgent as it has recently become because Joyce and his books were more hearsay than presence thanks to Irish censorship which from 1928 to the late 1960s was the most arbitrary and repressive in the western world. But Joyce's books are everywhere in Dublin "street furniture" (Joyce 1944: 211) he memorialized. Davy Byrne's, a drab pub in which one of the episodes of *Ulysses* is set has been all tarted up for the tourist trade. Bloom's house at 7 Eccles Street has been torn down but the door has been enshrined under glass in the Bailey restaurant across the street from Byrne's. And a new, or rather refurbished hotel in the heart of what Joyce headlined as "THE HIBERNIAN METROPOLIS" (Joyce 1986: 7.1) has been renamed Bloom's Hotel. This could be read as a sort of commercial revenge, Dublin getting its own back by exploiting the exploiter, but it can also be read as the extent to which (in paraphrase of an old Glasgow drinking song which Joyce expropriates [16.1164-65): "When I've had a couple of drinks of a Saturday, / Dublin belongs to me." The mythical city of Dublin belongs to Joyce and confirms his boasts (another chip on the shoulder) that if Dublin were to be destroyed it could be rebuilt from his books.

Many Dubliners, triggered by the arrogance of Joyce's wit, will tell you that that boast is simply not true. In apparent desperation one Dublin critic has even argued that Joyce is a poor writer because his autobiographical portrayal of himself as Stephen Dedalus in *A Portrait* is riddled with inaccuracies, exaggerations, and omissions. What kind of a writer is that: one who doesn't even have the memory and the verbal skill to tell his own life-story straight? Several Dubliners have complained that *A Portrait* and *Ulysses* are the autobiography of a braggart, that Joyce was bragging about his own skills when he described Stephen Dedalus's unbelievably facile composition of a villanelle in *A Portrait* (chapter 5:B) and when he describes Stephen's improvisation on Shakespeare's art and life during the discussion of *Hamlet* in the Scylla and Charybdis episode of *Ulysses,* an improvisation which is informed by a brilliant rodomontade of quotations from and allusions to Shakespeare's poems, many of the sonnets and all but two of his plays. Joyce, these critics argue, is portraying himself as even more brilliant than he actually was. In kindness to these defenseless critics I should add that the conviction that all of Joyce's works are fundamentally autobiographical has had and continues to have a very vigorous life among Irish men and women of letters. In extenuation: I am reminded of an Irish architect (presently in exile in Australia) I knew who complained that he had never been able to read *Ulysses* because for him the documentary realism with which Dublin and its citizens and the cadences of their voices was presented irrevocably compromised his sense of fiction; that is, it fractured the aesthetic distance he regarded as necessary to the experience of *Ulysses* as a novel.

I dwell on these crotchety Dublin responses not to mock them but because they suggest certain urgencies of personal and environmental presence that may cloud not only a Dubliner's reading but also, and at a distance, ours, urgencies that may account for the applicability to Joyce of the proposition. "A prophet is not without honor, save in his own country, and in his own house" (Matthew 13.57); and in Joyce's case the "Placeless cosmopolis" of that house may include, in Abraham Lincoln's delicious phrase, "this, his almost chosen people" (Lincoln 1861).

I wish now to address myself to the question: is Joyce a prophet? and to what extent does he share his mentor Ibsen's view (1865-66) that the artist must "rid [himself] of the aesthetic" in favor of "the ethical, the prophetic"? (Letter to Bjøorsen,. 12 September 1865, qtd. in Ibsen 1966: *Brand* 14) that the artist must, in Ibsen's words, "use the gifts God has given [him] to wake [his countrymen out of their lethargy, to see the direction in which the great problems of life are pointing"? (Letter to King Carl of Norway, 16 March 1866, qtd. in Ibsen 1964*)*.

To take *A Portrait* first (and out of the sequence of publication): in chapter 2:B, when Stephen is roughly ten years old, we are told that a "change of [his family's] fortune was reshaping the world about him into a vision of squalor and insincerity." He is described as chronicling "with patience what he saw, detaching himself from it and testing its mortifying flavor in secret." There follow three examples from that chronicle. The first two are visions of squalor and insincerity; the third is a description of a child's birthday party and a moment of childhood romance. "The next day" he tries to write verses on that suspended moment of romance:

> all these elements which he deemed common and insignificant fell out of the scene. There remained no trace of the tram itself nor of the trammen nor of the horses: nor did he and she appear vividly. The verses told only of the night and the balmy breeze and the maiden lustre of the moon.

The title Stephen awards to his poem, "To E— C—," suggests that it is in the manner of the early and callow Lord Byron. The clear implication is that "the aesthetic" has triumphed over and displaced what Joyce would have called the "mean and scrupulous style" of the other two scenes Stephen has chronicled.

In *A Portrait* chapter 5:B Stephens writes "verses for her again after ten years." He composes the villanelle I mentioned earlier, and during this second and more extended scene of artistic enterprise, his bent is revealed as overwhelmingly inclined toward "the aesthetic." He is literally awash in if not drowning in the avant-garde aesthetic flood waters of the 1890s. He imagines himself (in somewhat inflated terms) as "a priest of the eternal imagination, transmuting the daily bread of experience into the radiant body of everliving life." The villanelle he achieves is, if anything, even more utterly divorced from "the daily bread of experience" than "To E— C—" was, saturated with that blend of soft-religion, soft-sex and fruity Pateresque verbiage that so fascinated members of the Rhymers' Club and the decadent aesthetes of the 1890s. Stephen's villanelle is woven not from materials derived from "daily experience" but from the over-ripe aesthetic textures of other works of art.

My point is not that Stephen's villanelle is presented as a bad poem but rather that it is a *tour de force* indulgence in the narcissistic aestheticism which was, from Joyce's point of view, the dead end of nineteenth-century romanticism. Stephen's poem enables us to locate and evaluate him as artist. And if we recognize *A Portrait* not as autobiography but as the biography of an alter-ego, we can see that the alter-ego as mired in "the aesthetic" which the mature artist, in Ibsen's terms, must abandon in favor of "the ethical, the prophetic." In effect, in drawing the portrait of Stephen Dedalus in *A Portrait*, Joyce created an alter-ego in order to exorcise the ghost of the decadent aestheticism that haunted him and that haunted the arts in the late nineteenth and early twentieth centuries.

It is notable that Joyce himself, when he was Stephen's age, was so intensely interested in Ibsen that he taught himself Dano-Norwegian in order to be able to read Ibsen in the original. All that interest in and commitment to Ibsen is edited out of *A Portrait*, and, when Stephen announces at the end of the novel that he goes "to encounter for the millionth time the reality of experience and to forge in the smithy of my soul the uncreated conscience of my race," we may catch beneath the surface of the inflated prose an echo of the ethical and prophetic purpose of Ibsen, a purpose for which Stephen as zealous missionary is, at least as yet, poorly equipped. He is, as we are to discover in the opening pages of *Ulysses*, setting out on the wrong course, the flight not of Deadalus but of Icarus, due for a tumble, a drowned mission, a mouthful of seaweed.

Joyce said of *Dubliners* which was completed in 1907 but not published until 1914: "My intention was to write a chapter of the moral history of my country, and I chose Dublin for the scene because that city seemed to me the centre of paralysis" (qtd. in Gorman 1939: 150). This statement of intention is clearly ethical, prophetic, and the cumulative effect of the fifteen stories in *Dubliners* is that of an unrelenting condemnation of Irish paralysis and its root cause, simony (the prostitution by Dubliners of their moral and spiritual values in favor of a soul-deadening psychological and material security). The theme of simony-paralysis is unwittingly announced in the first paragraph of the first story, "The Sisters," by the boy-narrator of that story: "I said softly to myself the word paralysis. It had always sounded strangely in my ears, like the word gnomon in the Euclid and the word simony in the Catechism" (Joyce 1948: 22). To the boy it is the sounds that link the three words; to use it should be the sense, and since gnomon is not only a geometrical figure but also the shape of the pointer on sun dials, it drops out of the equation once it has pointed us to the relation between simony and paralysis.

Simony as self-prostitution figures subtly in all of the stories and climactically in the stories of public life, those that deal with images from politics, art, and religion, institutions that should nourish Irish life instead of paralyzing it. In "Ivy Day in the Committee Room" several down-at-the-heel ward heelers work for their candidate "Tricky Dicky Tierney" whom they disparage and whom we should despise (if we take the preferred hints). Their commitment is not to the candidate but to the dozen bottles of stout he sends around from his pub and to the money he has promised them. Because it is Ivy Day, the anniversary of the death of Charles Stewart Parnell, what they are really doing is re-enacting on the squalid level of ward politics in Dublin the "betrayal" of Parnell in Committee Room 15 of the House of Parliament at Westminster, an event that occurred twelve years before the time of the story. The giant figure of Parnell, symbolic of political energy and integrity, looms above the story to crush the inconsequence of the action in that committee room in Wicklow Street in the center of Dublin on a rainy October evening.

The story "Grace" focuses in a not dissimilar way on religion. The Jesuit priest who presides over the businessmen's retreat at the story's end styles himself as "a man of the world . . . their spiritual accountant" (Joyce 1948: 138). He is aptly (if crudely) named Father

Purdon so that we, or rather Joyce's Dublin contemporaries, wouldn't miss the point; Purdon Street (the name has since been changed) was in the heart of Dublin's flourishing red light district in 1900. Father Purdon seeks to reassure the businessmen in his audience by reducing spiritual commitment and integrity to something that sounds like double-entry bookkeeping, and as we exit from the story we realize that its title "Grace" does not mean what we might at first have expected, "a supernatural gift bestowed upon us by God for our salvation" (*Catechism* 1883: 24); it means instead the extra time allowed a debtor to enable him to settle his debts in the world of commerce.

Dubliners is a Jeremiad; as in the lamentations of Jeremiah, the prophet's voice utters a sequence of dirges, laments for what has happened to the chosen people in the promised land as a result of their spiritual transgressions: the land is desolate, the people are in bondage, their temple destroyed, the kingdom of Judah fallen. And as with the Israelites, so with the Irish: their paralyzed state is a just punishment for their simony, or, as the central character of the final story, "The Dead," half-realizes as he swoons toward sleep (death), the inability of Dubliners to "pass boldly into that other world in the full glory of some passion" (Joyce 1948: 173).

Many readers who experience *Dubliners* not just as a collection of individual stories hut as an integrated whole greater than the sum of its parts regard it as excessively dark and unrelenting, and Joyce himself came to feel that he may have been "unnecessarily harsh" (Ellmann 1982: 231). He therefore undertook the final story, "The Dead," to alleviate that harshness, though not with unqualified success. Recently a new group of psychological revisionists has come along to declarify the question of the harshness of *Dubliners* further by arguing that Joyce was unnecessarily harsh out of psychological necessity: he had to overstate the case in order to justify his desertion of his country and to expiate his guilt over his betrayal of his countrymen. But Joyce was hardly alone in his harshness, and we could, without much trouble, recruit a sizeable chorus of over-staters from the outstanding Irish writers of the last one hundred years. Perhaps we should remind ourselves of Yeats's great sentence about Swift, another Dubliner capable of harsh utterance: "Jonathan Swift made a soul for the gentlemen of this city by hating his neighbor as himself." I suspect that the weakness of *Dubliners* as an integrated whole does not lie so much in its unrelenting harshness as it does in the fact that the stories are insufficiently informed by ironic or suggestive implication of what the spiritual vitality would be that would release Dubliners from their houses of bondage, from the seemingly endless cycles of simony and paralysis.

In *Ulysses* no such want of information about the vitalizing principle should unsettle the reader, though it has not been fashionable in the Joyce industry to notice and fully deploy that information. The core of the ethical principle that underlies *Ulysses* is articulated, or rather fumbled out by that unlikely epic hero, Leopold Bloom, in the Cyclops episode. Homer described the Cyclops as "giants, louts, without a law to bless them" (Homer 1961: 160); they have no tradition of mutual help; they live instead in mutual indifference and scorn. And in Barney Kiernan's pub where the Cyclops episode of *Ulysses* is set, the atmosphere is redolent of mutual indifference and scorn; it is an atmosphere of macho-competitiveness in which manhood is established by swagger and violence and curse. Kiernan's pub is an all too appropriate setting because in "real life" Kiernan was a collector of souvenirs of violent crime, and his pub featured changing displays of these curios.

The story of the Cyclops episode is told by an anonymous barfly who is a master of colloquial cynicism and scorn. His story is interrupted from time to time by parodies intruded by a stage-manager whose technique Joyce described as gigantism. In one of these parodies Kiernan's pub is consecrated as Dublin's permanent cathedral. Catholic Dublin has been resentfully housed in a pro-Cathedral since the reformation took over the great medieval Cathedrals of Christ Church and Saint Patrick's. Kiernan's pub as Cathedral enshrines force and violence as the ethical principles that make men men and that make nations nations, the principles that will make Ireland, in the words of Thomas Osborne Davis, "a nation once again."

Bloom objects that if Ireland realizes itself as a nation through force and violence, it will persecute not only others but also its own people in the name of "discipline." "Wouldn't it be the same here," he asks, "if you put force against force?" (Joyce 1986: 12.1361). And the answer which is supposed to put Bloom down is pounded into the table by the number-one-boy Cyclops and unanimously carried: "We'll put force against force" (12.1364).

A few pages later Bloom cites the persecution of the Jews in Morocco as further evidence of man's inhumanity to man and is asked in anti-Semitic scorn:

> –Are you talking about the new Jerusalem? says the citizen.
> –I'm talking about injustice, says Bloom.
> –Right, says John Wyse. Stand up to it then with force like men. (2.1473-75)

At this point the story-teller intrudes his utter contempt for Bloom's unmanly behavior and says that Bloom "collapses . . . as limp as a wet rag" (12.1479-80). But Bloom continues (bravely, considering the context):

> –But it's no use . . . Force, hatred, history, all that. That's not life for men and women, insult and hatred. And everybody knows that it's the very opposite of that that is really life.
> –What? says Alf.
> –Love, says Bloom. I mean the opposite of hatred. (12.1481-85)

At that moment Bloom interrupts himself to leave the pub on an errand and is dismissed with heavy sarcasm by the otherwise anonymous "citizen" as "A new apostle to the gentiles" (12.1489) and buried beneath a wave of anti-Semitic remarks. In effect Joyce puts the imprimatur on Bloom's far-from-glib statement of principle by the devastating irony leveled at Bloom's presumably Christian opponents, an irony that climaxes when Bloom's chief opponent, the "citizen," shouts, "By Jesus, I'll crucify him so I will" (12.1812) and heaves an empty biscuit tin at him. At the end of the episode in a parody of Biblical prose Bloom is mocked as ben Bloom Elijah being assumed bodily into heaven; and the final delicate irony that shimmers behind the parody and its mockery of Bloom is that Bloom is, unbeknownst to himself and to everyone except his Maker, Elijah-incognito as promised by the prophet Malachi at the end of the Old Testament:

> Behold. I will send you Elijah the prophet before the coming of the great and dreadful day of the Lord: And he shall turn the heart of the fathers to the children, and the heart of the children to their fathers, lest I come and smite the earth with a curse (4. 5-6).

An elaborate (and cursory) premonition of Bloom's statement of ethical principle is woven into the Sirens episode which precedes Cyclops. The Sirens episode takes place in the restaurant and bar of the Ormond Hotel in central Dublin. In 1904 the Ormond was to amateur musicians what Kiernan's pub was to crime freaks. The episode is cast in the form, Joyce said, of a *fuga per canonem,* a fugue according to the rules, and I take it that the subject/counter-subject of the fugue is love and war, announced by the duet "Love and War"

(11.459 ff.) and structurally confirmed by the two halves of the episode. The first half is dominated by a song of love and loss; the second half by a ballad of war and loss. Between the halves Blazes Boylan, Bloom's brash and aggressive rival for Molly Bloom's bed if not for her affections, arrives at Bloom's house in Eccles Street for the rendezvous he and Molly have arranged. Earlier in the fugue Boylan has entered as "conquering hero" (11.340); Bloom as "unconquered hero" (11.342). Boylan will conquer; Bloom will not be conquered. How Bloom will respond to being cuckolded and remain unconquered we do not know at this point in the novel; nor does he, though he does know intuitively and accurately that he will be cuckolded. From behind the scene Joyce, or the invisible concertmaster weaving the fugue together, lets us know what in principle Bloom's response should be.

Stephen's father Simon Dedalus has just finished singing Lionel's lament for his lost love, Martha, from Flotow's opera *Martha,* a song ironically appropriate to Bloom's impending loss. In the entr'acte that follows several things happen thanks to the conductor's baton. Father Cowley, the pianist who has accompanied Simon Dedalus, plays the "Minuet of *Don Giovanni*" (11.965) which occurs on stage in the opera while off-stage the Don is attempting to seduce the young peasant woman Zerlina and her fiancé Massetto is being distracted by Leporello, the Don's valet. The parallels to the situation in *Ulysses* are obvious except that Don Giovanni's attempted seduction fails while Boylan's succeeds. A third of a page later in *Ulysses* Boylan arrives at 7 Eccles Street and the sound of his knock on the door enters the fugue. Immediately the pianist suggests as the song for the bass who is about to sing "*Qui sdegao*" (11.990)—"Here indignation . . ." —an Italian version of an aria from Mozart's opera *The Magic Flute*. (The libretto of the opera is in German, but it was frequently sung in Italian in the nineteenth century because, as everybody knew, Italian was opera-language.) The aria, *In diesen heil'gen hallen* ("In these sacred halls") is sung by Sarastro, the High Priest of Isis and Osiris in Act 2, Scene 3 of the opera. Earlier in that scene the Queen of the Night in her revenge Aria has demanded that her daughter murder Sarastro or be disowned. When Sarastro enters, the Queen of the Night's daughter instead of murdering him begs his forgiveness. His aria answers her plea. He explains that "In these sacred halls" of the Temple of Light over which he presides there is no such thing as revenge because only love can bind human beings together.

What has been inserted here is a super-cryptic ethical reminder that indignation-confrontation-revenge (the domestic corollary to the political force and violence advocated in Cyclops) are not, as they are conventionally assumed to be, the answer to the sort of marital dilemma Molly and Boylan present to Bloom. The answer, as Bloom later defines it, is "Love . . . the opposite of hatred" (Joyce 1986: 122.1485). But the pianist's suggestion and the ethical sign-post are just two Italian words, swept aside in favor of the bass's rendering of "The Croppy Boy," a ballad of the Rebellion of 1798 in which the Irish Croppy Boy whose father and brothers have died in the Rebellion goes to his own death when he confesses to a priest who is an English sergeant in disguise. At the violent climax of the ballad fragments of sound enter the fugue to suggest that Boylan is entrenched at 7 Eccles Street; the sergeant in the ballad throws off his disguise, and the ballad continues: "I hold this house. Amen. He gnashed in fury" (11.1120). As, in the fury of his lust, Boylan holds Bloom's house.

The resolution of the marital dilemma towards which Bloom moves in the course of this day of preoccupation has several strands. One of those strands is a growing sense of his own complicity in Molly's behavior. He comes to realize that, as Stephen says of Shakespeare and Anne Hathaway's "adulterous behavior": it was a sin "committed by another in whose sin he too has sinned" (9.1009). Stephen's remark implies that as in original sin, "we were all made partakers of the sin and punishment of our first parents" (*Catechism* 1883), so in adultery both husband and wife have sinned, and this Bloom comes to realize (although "sin" is not a word that fits comfortably in Bloom's vocabulary).

Another strand in Bloom's resolution of the marital dilemma derives from one of the fundamental ways in which he differs from his fellow Dubliners and their pub culture. Several times in the course of the day Bloom wonders why, sixteen years ago, Molly had chosen him. During the evening he gets half an answer when he recalls Molly saying: "Because you were so foreign from the others" (13.1209-10). A page from the end of the novel Molly gives a somewhat fuller answer: "yes that was why I liked him because I saw he understood or felt what a woman is" (18.1578-79). And the stage manager of the novel has gone to considerable lengths to demonstrate the validity of Molly's perception by arranging coincidences and overlapping passages that demonstrate Bloom's extraordinary capacity for intuitive sympathy with and insight into the inner lives of others, both men and women, and particularly on this day of crisis, women and Molly. At one climactic moment of hallucination Bloom is mocked as "the new womanly man" (15.1798-99), but throughout the novel Joyce is busy affirming what he would have regarded as Bloom's womanly traits, his intuitive awareness of others and his capacity to care for, to nurture and heal others. The patron saint of Odysseus, Bloom's epic counterpart, was Pallas Athena, goddess of mother wit and wisdom and of the household arts. It is no accident of craven uxoriousness that when we first meet Bloom in the novel he is preparing breakfast for himself and Molly, under the aegis of his nameless Pallas Athena, no doubt. (And how the Dublin barflies would have laughed at that lack of manhood—and invited Athena's silent retaliation.)

Now I come to the principal strand in Bloom's resolution of his dilemma: his voyage toward himself as "a conscious reactor against the void of incertitude' (17.2210-11), a self which can accept and "justify to himself his sentiments" (17.2211), his "equanimity" (17.2177), with the realization that given "natured nature" (17.2179), the world of actualities, one must learn to accept with equanimity "the futility of triumph or protest or vindication: the inanity of extolled virtue: the lethargy of nescient matter: the apathy of the stars" (17.2224-26). And in this case apathy is not just indifference but the Stoic discipline of *apathia*, the mastery of the passions through the ascendancy of reason.

The third strand develops into the biggest joke at Dublin's expense, one of the major components of that chip on Joyce's shoulder. On occasion in *Ulysses* we are reminded that it is as if Bloom were playing the prophetic part of Elijah, but throughout the novel he is subtly portrayed as the best and probably the only "Christian" in Catholic Christian Dublin. On the surface this "anonymous Christian" (to borrow Karl Rahner's phrase) is a Jew who knows little of the religion of his ancestors, and whose experience of the institutions of Christianity seems to have been limited to the fact that he was baptized twice, once "in the protestant church of Saint Nicholas Without" (17.543) and once in the Catholic "church of the Three Patrons" (17.546).[3]

In the Hades episode at eleven in the morning Bloom attends his acquaintance Paddy Dignam's funeral. In the course of the funeral one of the mourners is taking up a collection to tide the dead man's family over until his insurance can be settled. We learn of this collection when Bloom is out of earshot. In mid-afternoon, 118 pages and four episodes later in the day, we learn from the mourner who took the collection that Bloom pledged and contributed to the fund while the mourners were still in the cemetery. The scene in which Bloom does so is left out of the Hades episode to indicate that it was not self-consciously present in Bloom's experience, that it was an act of unselfconscious generosity, a true act of Christian charity, *caritas*, love given freely, not given in the hope of spiritual or material gain, nor out of fear of divine retribution nor in hope of achieving possession of another.[4]

Bloom's *caritas* is first encoded in this omission of a scene from the Hades episode. Later in the novel the tactic of omission is repeated. An entire episode is omitted from Bloom's day, an episode that takes place between six and eight o'clock in the evening when Bloom and three of his acquaintances visit the newly widowed Mrs. Dignam to comfort her,[5] to give her the money that has been collected during the day, and to try to sort out the tangled question of her late husband's mortgaged insurance. Bloom is invited by the others because he once worked for an insurance company. He doesn't recall the episode itself, only that it was fatiguing, and in the final review of his day, the omitted episode is summed up as "a blank period of time including a cardrive, a visit to a house of mourning, a leave-taking (wilderness)" (17.2051-52), another act of unselfconscious and even painful generosity, *caritas*.

St. Augustine in his commentary on St. Paul's praise of charity in I Corinthians 13 sets charity (*caritas*) in opposition to cupidity (*cupiditas*). In Dublin the typical anti-Semitic caricature identified the Jews with cupidity, a motif that is played throughout *Ulysses* and focused, of course, on Bloom, The irony of that focus cuts two ways: against Christian Dublin and its violent and scornful lack of *caritas* and in favor of Bloom, the lapsed Jew who, alone in Dublin, is consistently capable of the intuitive, sympathetic concern for others which is the stuff of *caritas*. St. Augustine also remarks in passing that *caritas* is most difficult to achieve and sustain in relation to those with whom one is most intimate. Bloom's realization of his own complicity in Molly's "sin" and the "equanimity" of acceptance with which he resolves this day of crisis (the crisis of middle life) imply but omit to state that Bloom has forgiven Molly. Once again the omission confirms the act of forgiveness as *caritas*, not the glib repetition of the formula that is supposed to govern the just and charitable man ("as we forgive those who trespass against us") but the living deed of forgiveness itself. This *caritas* echoes back and is confirmed by what the concertmaster has encoded as ethical principle in the Sirens episode: in these sacred halls there is no such thing as revenge, only love can bind human beings together. And we may well be reminded of Reinhold Niebuhr's dictum: "Therefore we must be saved by the final form of love, which is forgiveness."

There is further encoded confirmation of Bloom's "anonymous Christianity." Twice after midnight Bloom's day is reviewed, once in the form of a Catholic litany (15.1940-52) and once in the form of a Jewish litany (17.2044- 58). The suggestion is that Bloom's day is informed by liturgical significance. When the catechetical questioner in the Ithaca episode asks: "Had he performed any corporal work of mercy for her?" (17.487) the identity of the "her" doesn't matter, but the opportunity for an immediate answer in the text should preclude a broader application of the question. Each morning devout Dublin Catholics should, as instructed by the *Maynooth Catechism*, pray to be able to perform one or more of the seven corporal and seven spiritual works of mercy in the course of the day. A review of Bloom's day reveals that he has managed unselfconsciously to perform all fourteen works of mercy in the course of the day.

Seven corporal works of mercy:

(1) to bury the dead. (Bloom has attended Dignam's funeral.)
(2) to clothe the naked. (Bloom has sent a hat to his daughter Molly on the occasion of her birthday.)
(3) to feed the hungry. (His wife, his cat, the gulls on the Liffey, an unidentified dog, and he at least attempts to feed Stephen.)
(4) to give drink to the thirsty. (The cat, Molly, and Stephen.)
(5) to harbor the houseless. (Stephen is houseless at the end of the day, and Bloom offers him shelter.)
(6) to visit those in prison. (Dublin is, as Bloom repeatedly reflects, "a house of bondage" [7.209].)
(7) to administer to the sick. (Stephen, drunk, is knocked out in a street brawl, and Bloom stands by to help him recover.)

Seven spiritual works of mercy:

(1) to admonish sinners. (Bloom reproves "the citizen" for his anti-Semitism in the Cyclops episode, and he admonishes Stephen about drinking too much and about wasting his substance with whores.)

(2) to bear wrongs patiently. (What else?)
(3) to comfort the afflicted. (Josie Breen who is saddled with a dotty husband, the widow Dignam, and Stephen.)
(4) to counsel the doubtful. (Stephen, drunk and boasting, is superstitiously frightened by a clap of thunder and begins to doubt that he is still in "the land of Phenomenon" [14.439]—Bloom assures him that he is.)
(5) to forgive offenses. (What else? At least Bloom has numerous opportunities and only blows one, at the end of the Cyclops episode.)
(6) to instruct the ignorant. (In one review of Bloom's day he is called "mentor of Menton" [15.1943) who is ignorant of "a dinge in the side of his hat" [16.1015].)
(7) to pray for the living and the dead. (Bloom attends a Mass in Lotus Eaters and listens while the priest prays in English at the end [15.420-447]. He hears prayers for the soul of Paddy Dignam in the Hades episode, and in ten days he will make his eighteenth annual pilgrimage to visit his father's grave.)

Each day devout Dublin Catholics also prayed to be made whole in the seven virtues: Faith, Hope, Charity, Prudence, Justice, Fortitude, Temperance; and a review of the novel confirms that in subtle ways Bloom's capacity for each of these virtues is whole and intact. His detractors repeatedly emphasize this by calling him "the prudent member" (Joyce 1986: 12.211) in scorn and by assailing him for his temperance (an odd-man-out virtue in Dublin's pub culture).

But Bloom is not presented as infallible. When he is about to enter Bella Cohen's brothel in the course of his Good Samaritan mission to see Stephen through a drunken evening, Bloom trips on the stairs. Superstition would say that he is reluctant to enter. The young prostitute who is guiding him says "Don't fall upstairs" (15.2025). In reply Bloom quotes proverbs 24.16: "The just man falls seven times" (15.2027). That passage is traditionally glossed: the just or virtuous man will have at least one brush with each of the seven deadly sins (pride, lust, envy, anger, covetousness, gluttony, and sloth) in the course of his life, and—sure enough—the grocer's assistant's mind that is stocking the shelves in this novel has arranged for Bloom to have encounters with all seven.

Dubliners are fond of suggesting that their Catholic country is the most Christian country in the world. The joke at their expense:

Bloom, the heterodox Jew-Protestant-Catholic-Freemason-Agnostic-Atheist is the only practicing, unselfconsciously virtuous (and sinful) "anonymous Christian" in Dublin.

In conversations reported by a young Irish friend named Arthur Power, Joyce described what he was attempting in *Ulysses* as "the new realism." Power neglected to ask the follow-up question: what is the nature of the "reality" on which the new realism is based? But at another point in the conversations Joyce identified that reality. He was talking about the Circe episode which is dominated by nine extended hallucinations or nightmares, and he said that in that episode "I approached reality closer in my opinion than anywhere else in the book except perhaps for moments in the last chapter [the Penelope episode]. Sensation is our object, heightened even to the point of hallucination" (Power 1974: 75). Reality, in other words, is not just fact or probability but "sensation . . . hallucination," and those psychological realities comprise the "object" of the new realism both in the sense of the thing observed and in the sense of the writer's goal because that goal is, Joyce said, to bring into focus "the subtitles, equivocations and the subterranean complexities which dominate the average man and comprise his life" (Power 1974:74).

At another point in the conversations, Joyce suggested another dimension of the ethical purpose of the new realism: it seeks to achieve "the sudden reality which smashes romanticism to a pulp" (Power 1974: 98), and by romanticism Joyce meant at best abstract and idealized emotion; at worst, emotion prettified, faked, sentimentalized. In Joyce's view, romanticism and the idealisms it spawns lead human beings to delude themselves and to impose upon and persecute others, and the answer to romanticism was not the composure and restraint of the "classicism" many of his contemporaries were advocating but the crushing force of "the new realism."

What I have been attempting to outline are some of the ethical, the prophetic components of the chip on Joyce's shoulder, the challenge Joyce presents to his countrymen and to us. But there are so many more components to be explored. Among them there is that fabulous technical skill and exuberance with language which most of us first encounter as the sort of difficulty that makes us knot our brows and clench our pencils. There is also that unremitting and courageous refusal to compromise commercially or artistically (even when close friends and competent critics complained of many of the later episodes of *Ulysses* and of the whole project of *Finnegans Wake* when it was appearing serially as *Work in Progress*. And there is the challenge of the sheer scale of *Ulysses* and *Finnegans Wake* as though both were comprised of six or seven major novels the size and weight of Dickens's or George Eliot's or Henry James's fictions, as though those severally had been layered into one, not so much in the linear, horizontal flow of narrative as in vertical striation, strata above strata until *Ulysses* becomes at once the drama of a day and the epic of the middle course of life; *Finnegans Wake*, the drama of a night and a panoramic vision of the mythopoeic space which we know as human history.

Samuel Beckett characterized Joyce as an "Apollonian" artist and said "he's tending toward omniscience and omnipotence." What Beckett suggests is that Joyce was not so much "breaking new ground" (as the Danish critic Georg Brandes had said of Ibsen), but preempting the ground, achieving an Apollonian inclusiveness which is more conclusion than it is beginning, much what Samuel Johnson said of Shakespeare, what Matthew Arnold implies in the opening line of his sonnet to Shakespeare: "Others abide our question. Thou art free" (Arnold 32). Except that I'm reminded of a story of Otto Luening's about an encounter with Joyce in Paris in the 1920s. Otto sought to call himself to the great man's attention by saying that he was a freethinker, and Joyce replied: "My dear young man, very few of us can think and none of us is free!"

Coda: When Joyce looked in the mirror whom did he see besides that grocer's assistant possessed of what he called "a meticulousity bordering on the insane"? One answer lies encoded in an essay, "Irish Books" by the Irish man of letters W. K. Magee (pseudonym, John Eglinton, whom Yeats praised as "the only Irish critic of his generation").

> If a masterpiece should still come of this literary movement, we would not be surprised if it appears by a kind of accident and in some unexpected quarter, and we have a fancy that appearances in modern Ireland point to a writer of the type of Cervantes rather than to an idealizing poet or romance writer. A hero as loveable as the Great Knight of the Rueful Countenance might be conceived, who in some back street of Dublin had addled his brains with brooding over Ireland's wrongs, and that extensive but not always quite sincere literature which expresses the resentment of her sons toward the stranger [i.e., the English overlords]. His library would he described, the books which had 'addled the poor gentleman's brains with brooding over Ireland's wrongs' . . . We can conceive him issuing forth, fresh-hearted as a child . . . to realise and incidentally to expose the ideals of present-day Ireland, What scenes might not be devised at village inns [or Dublin pubs] . . . what blanketings, in which our sympathies would be wholly with the rebel

> against the despotism of fact! . . . And such a book . . . need not really insult . . . the cause of Irish nationality any more then Cervantes laughed real chivalry away. (Magee 1917: 87-89)

In *Ulysses* during the scene in the National Library when Stephen is holding forth on Shakespeare, John Eglinton is present as a character in the fiction, and when this passage in his essay is alluded to, Stephen is listening. Everyone else in the room is talking at once, and someone says "Moore is the man for it" (9.309-10); that is, the Irish-born novelist and playwright George Moore will be the Irish Cervantes. Fair enough, except that when you hold *Ulysses* in your hand, what you are holding is the comic national epic John Eglinton has described: right down to the detailed list of the books in Bloom's library (17.1362-1407). The equation reads: as Bloom is an Irish Don Quixote, so Joyce is the Irish Cervantes, and *Ulysses* is to Ireland as *Don Quixote* is to Renaissance Spain, what the Homeric poems are to ancient Greece. As Joyce would say, "How's that for high?" (Joyce 1966: I: 104): or as a Dubliner might say: "How's that for a chip on the shoulder?"

Another: in the midst of a laundry list of Irish heroes in the Cyclops episode appears the name Patrick W. Shakespeare (Joyce 1986: 12:190-91). During the library scene John Eglinton asks, "Has no one made him [Shakespeare] out to be an Irishman?" (9.519-20). He then adds that an Irish judge is "searching for some clues" (9.520) and mentions the fact that Hamlet swears "by Saint Patrick (9.521). But what really accounts for Patrick W. Shakespeare as Irish hero is the substitution of a name. Another encoded chip: one of the outstanding social and cultural historians of Ireland at the beginning of this century was Patrick Weston Joyce (1827-1914) (no relation to JJ). He lends part of his name not to the assertion that Shakespeare was Irish, but to Joyce's modest assertion? "I am Ireland's Shakespeare."

And one for the road: toward the end of her soliloquy and the end of the novel, Molly Bloom experiences a seizure of exasperation at the onset of her menstrual period and calls on her Maker for release. She doesn't say "O Lord, deliver me" or "God help us" she says "Oh Jamesy let me up out of this" (18.1128-29). She calls, in a rather familiar way, on her Maker, the novelist James Joyce, God in this creation.

Notes

1 References to *Ulysses* are to episode and line numbers of "The Corrected Text," ed. Hans Walter Gabler, Wolthard Steppe, and Claus Meichior (New York: Random House, 1986).

2 In discussing *A Portrait* I have found it useful to emphasize the episodic nature of the novel by focusing on its sub-chapters: chapter 1:A, chapter 1:B, etc.

3 The assertion that Bloom is a Jew needs to be qualified. He so regards himself and is so regarded by Dublin, but in Jewish tradition religious heritage is matrilineal, and Bloom's, mother was Ellen Higgins, second daughter of Julius Higgins (born Karoly) and Fanny Higgins (born Hegarty) (Joyce 1986: 17.536-37). Hegarty and Higgins suggest an Irish matrilineage. Molly, whose father was apparently Irish Catholic, knows little (or says little) about her mother Lunita Laredo though she does remark "on account of my being jewess looking after my mother" (18.184-85). I have no proof that Joyce knew about this tradition of matrilinearity, but whether he did or not it provides an opportunity for his kind of joke: that Molly is Jewish by heritage and Bloom is not.

4 In what I would argue is a misguided emendation in "The Corrected Text" of *Ulysses* (Joyce 1986: 9.429-31) phrases from Saint Thomas Aquinas's distinction between "true love" and "self love" drift through Stephens mind, both defining the sort of *caritas* Bloom unselfconsciously practices and (unfortunately) implying that Stephen is capable of love (or at least of knowing that the capacity for love is what he needs) which, as of 16 June 1904, he certainly is not. "Pain, that was not yet the pain of love, fretted his heart" the unidentified narrator remarks in a rare evaluatory intrusion (1.102).

5 The Dubliners Bloom encounters in the course of the day can hardly be called his "friends," they know so little of him and pay such cursory attention to him.

Works Cited

All Things Considered. 1995. National Public Radio. 27 October 1995.

American Heritage Dictionary of the English Language 1992. Boston: Houghton Mifflin.

Aristotle. 1967. *Poetics.* Trans. Gerald F. Else. Ann Arbor: University of Michigan Press.

Arnold, Matthew. 1954. "Shakespeare." *Matthew Arnold Poetry and Prose.* Ed. John Bryson. London: Rupert Hart-Davis. 32.

Augustine, St. 1963. *The Confessions.* Trans. Rex Warner. New York: Mentor.

Baym, Nina et al, eds. 1989. *The Norton Anthology of American Literature.* Third Edition. New York: Norton.

Bible, Holy. 2002. Authorized Version. Uhrichsville, OH: Barbour.

Blake, William. 1956. "Preface." *Milton: A Poem in Two Books. The Portable Blake.* Ed. Alfred Kazin. New York: Penguin Books. 411-44.

Bloom, Allan. 1987. *The Closing of the American Mind.* New York: Simon and Schuster.

Bone, Edith. 1957. *7 Years' Solitary.* New York: Harcourt Brace.

Borges, Jorge Luis. 1964. "The Circular Ruins." Trans. James E. Irby. *Labyrinths*

---. 1964. "Funes the Memorious." Trans. James E. Irby. *Labyrinths.* 59-66

---. 1964. *Labyrinths: Selected Stories and Other Writings.* Ed. Donald A. Yates and James E. Irby. New York: New Directions.

Bradshaw, John. 1990. *Homecoming: Reclaiming and Championing Your Inner Child.* New York: Bantam Books.

Britton, James. 1992. *Language and Learning.* 1970 2nd edition London: Penguin Books.

Brown, Roger. 1970. *Psycholinguistics: Selected Papers.* New York: Free Press.

Buñuel, Luis. 1983. *My Last Sigh.* Trans. Abigail Israel. New York: Knopf.

Bruner, Jerome. 1990. *Acts of Meaning.* Cambridge: Harvard University Press.

The Catechism Ordered by the National Synod of Maynooth. 1883. Dublin: n. p.

Chace, William. 1980. *Lionel Trilling: Criticism and Politics.* Stanford: Stanford University Press.

Chatwin, Bruce. 1988. *The Songlines.* 1987 New York; rpt.

Coleridge, Samuel Taylor. 1985. "Kubla Khan: Or, A Vision in a Dream." *Samuel Taylor Coleridge.* Ed. H. J. Jackson. New York: Oxford University Press. 102-104.

---. 1951. "Shakespeare's Judgment." *Selected Poetry and Prose of Coleridge.* Ed. Donald A. Stauffer. New York: Modern Library. 429-33.

Dante Alighieri. "The Letter to Can Grande" (*Epistolam X ad Canem Grandem della Scala*) in *Literary Criticism of Dante Alighieri.* Trans. and ed. Robert S. Heller. Lincoln, NB: University of Nebrfaska Press. 95-111.

Darwin, Charles. 1959. *The Autobiography of Charles Darwin: 1809-1882.* Ed. Nora Barlow. New York: Harcourt, Brace.

Dawidowicz, Lucy S., ed. 1976. *A Holocaust Reader.* New York: Behrman House.

DeCasper, Anthony. 1990. Unpub. Paper presented at the *American Psychological Association*, University of North Carolina at Greensboro, August 1990.
Dewey, John. 1934. *Art as Experience*. New York: Milton, Balch & Company.
Dickinson, Emily. 1970. *The Complete Poems of Emily Dickinson*. Ed. Thomas H. Johnson. London: Faber & Faber.
---. 1986. *Emily Dickinson: Selected Letters*. Ed. Thomas H. Johnson. Cambridge: Harvard University Press.
Edelman, Gerald M. 1992. *Bright Air, Brilliant Fire: On the Matter of the Mind*. New York: Basic Books.
Editorial. 1995. *Berkshire Eagle* 23 October1995. A6.
Ellmann, Richard. 1982. *James Joyce*. 1959 New York: Oxford University Press.
Emerson, Ralph Waldo. 1996. "Brahma." *Emerson: Essays and Poems*. Ed. Joel Porte. New York: Library of America. 1209.
Encyclopedea Brittanica. 1910-1911. 11th ed. 29 vol. New York: Cambridge University Press.
Ferris, Timothy. 1993. *The Universe and the Eye*. San Francisco: Chronicle Books.
Fischer, David Hacket. 1994. *Paul Revere's Ride*. New York: Oxford University Press.
Ford, Ford Madox. 1951. *The Good Soldier*. 1915 New York: Alfred A. Knopf.
---. 1950. *Parade's End*. New York: Knopf.
Freud, Sigmund. 1938. *The Interpretation of Dreams*. In *The Basic Writings of Sigmund Freud*. Trans. A. A. Brill. New York: Random House.
Fukuyama, Francis. 1989. "The End of History" in *National Interest* 16 (Summer 1989): 3-18.
Gifford, Don. 1990. *The Farther Shore: A Natural History of Perception, 1798-1984*. New York: Atlantic Monthly Press.
Gifford, Don, with Robert J. Seidman. 1988. Ulysses *Annotated: Notes for James Joyce's* Ulysses. 2nd ed. Berkeley: University of California Press.
Gide, André. 1973. *The Counterfeiters*. Trans. Dorothy Bussy. New York: Anchor Books.
Gorman, Herbert. 1939. *James Joyce*. New York: Farrar.
Gould, Stephen Jay. 1995. Review of George Johnson, *Fire in the Mind; Science, Faith and the Search for Order*. New York: Alfred A. Knopf, 1995 rpt. in *The Berkshire Eagle*, 19 October 1995, C5.
Graham-Dixon, Andrew. 1996. *A History of British Art*. London: BBC Books.
Graves, Robert. 1972. *The White Goddess: A Historical Grammar of Poetic Myth*. New York: Octagon Books.
Hawking, Stephen. 1996. *A Brief History of Time*. 1988. New York: Bantam.
Homer. 1974. *Iliad*. Trans. Robert Fitzgerald. New York: Doubleday.
---. 1961. *The Odyssey*. Trans. Robert Fitzgerald. Garden City: Doubleday.
Hopkins, Gerard Manley. 1986. "To What Serves Mortal Beauty?" *Gerard Manley Hopkins*. Ed. Catherine Phillips. New York: Oxford University Press. 167.
Hughes, Robert. 1993. "The Medium Inquisitor." Review of *The Collected Essays and Criticism of Clement Greenberg*. Vol. 1-4. Ed. John O'Brian in *New York Review of Books* 21 October 1993.
Ibsen, Henrik. 1966. *Brand*. Ed. and trans. G. M. Gathorne-Hardy. Seattle: University of Washington Press.
James, Henry. 1956 ."The Art of Fiction" 1884. *The Future of the Novel: Essays on the Art of Fiction*. Ed. Leon Edel. New York: Vintage Books, 3-27.
---. 1958. Letter to H. G. Wells (10 July 1915). *Henry James and H. G. Wells: A Record of Their Friendship, Their Debate on the Art of Fiction, and Their*

Quarrel. Ed. Leon Edel and Gordon N. Ray. Urbana: University of Illinois Press. 265-68.
James. William. 1898. *Human Immortality: Two Supposed Objections to the Doctrine*. Boston: Houghton, Mifflin.
---. 1981. *The Principles of Psychology*. 1890 3 vol. Cambridge, Harvard University Press.
Jaynes, Julian. 1976. *The Origin of Consciousness in the Breakdown of the Bicameral Mind*. Boston: Houghton-Mifflin.
Johnson, Samuel. 1963. *Johnson's Dictionary: A Modern Selection*. Ed. E. L. McAdam, Jr. and George Milne. New York: Pantheon.
Jones, David. 1952. *Anathemata: Fragments of an Attempted Writing*. New York: Chilmark Press, n.d.
---. 1978. *In Parenthesis*. 1937. New York: Faber & Faber.
Joyce, James. 1964. "Daniel Defoe (II)." Trans. Joseph Prescott in *Buffalo Studies* 1.1 (1964): 14-27.
---. 1948. *Dubliners*. 1914. *The Essential James Joyce*. Ed. Harry Levin. London: Panther. 22-29.
---. 1939. *Finnegans Wake*. New York: Random House.
---. 1966. *Letters of James Joyce*. Ed. Stuart Gilbert and Richard Ellmann. New York: Viking.
---. 1976. *A Portrait of the Artist as a Young Man*. 1916. Harmondsworth: Penguin.
---. 1975 *Selected Letters of James Joyce*. Ed. Richard Ellman. New York: Viking.
---. 1964. *Speeches and Letters*. Ed. Evert Sprinchorn. New York: Hill & Wang.
---. 1944. *Stephen Hero*. Ed. Theodore Spencer, John J. Slocum, and Herbert Caihoon. New York: New Directions.
---. 1986. *Ulysses*. 1922. "The Corrected Text." Ed. Hans Walter Gabler, Wolfhard Steppe, and Claus Melchior. New York: Random House.
Keller, Helen. 1974. *The Story of My Life*. 1902. New York: Dell.
Kenner, Hugh. 1968. *The Counterfeiters*. Bloomington: Indiana University Press.
Langer, Suzanne. 1951. *Philosophy in a New Key: A Study in the Symbolism of Reason, Rite and Art*. 1942. Cambridge: Harvard University Press.
Lessing, Gotthold Ephraim. 1962. *Laocoon*. Trans. Edward Allen McCormick. Indianapolis and New York: Bobbs-Merrill.
Levi, Primo. 1984. *The Periodic Table*. 1975. Trans. Raymond Rosenthal. New York: Shocken Books.
---. 1961. *Survival in Auschwitz*. Trans. Stuart Woolf. 1959. New York: Collier Books.
Lincoln, Abraham. 1861. To the New Jersey State Senate, 21 February1861.
Locke, John. 1986. "From *An Essay on Human Understanding*." (1690). *The Norton Anthology of English Literature*. Ed. M. H. Abrams, et al. Fifth Edition. Vol. 1. New York: Norton. 1752-57.
Maddock, Kenneth. 1972. *The Australian Aborigines: A Portrait of Their Society*. London: Allen Lane.
Magee, William Kirkpatrick (John Eglinton). 1917. "Irish Books." *Anglo-Irish Essays*. Dublin: Talbot.
Melville, Herman. 1983. *Moby Dick. Redburn, White Jacket, Moby Dick*. New York: Library of America.
Milton, John. 1953. "Lycidas." *Poems of John Milton*. 141-51.
---. 1953. *Paradise Lost. Poems of John Milton*. 205-487.
---. *The Poems of John Milton*. Ed. James Holly Hanford. 2nd ed. New York: Ronald Press.

Morrison, Toni. 2006. "Nobel Lecture." http://nobelprize.org 7 August 2006 10:13a.m.

The New York Times, 6 April 1993, C11.

The New York Times, February 1995. 59.

The New York Times, 30 May 1995, C1.

Nussbaum, Martha. 1986. *The Fragility of Goodness: Luck and Ethics in Greek Tragedy and Philosophy*. Cambridge: Cambridge University Press.

Park, Clara. 1991. *Rejoining the Common Reader: Essays, 1962-1990*. Evanston, Illinois: Northwestern University Press.

Pfeiffer, John E. 1982. *The Creative Explosion: an Inquiry into the Origins of Art and Religion*. New York: Harper and Row.

Pinker, Steven. 1994. *The Language Instinct*. New York: William Morrow.

Pope, Alexander. 1993. "An Essay on Criticism." *Alexander Pope: A Critical Edition of the Major Works*. Ed. Pat Rogers. New York: Oxford University Press. 17-39.

Power, Arthur. 1974. *Conversations with James Joyce*. New York: Barnes and Noble.

---. 1949. *From an Old Waterford House*. London: Mellifont Press.

Pyne, Stephen J. 1991. *Burning Bush: A Fire History of Australia*. New York: Henry Holt.

Rosenfield, Israel. 1988. *The Invention of Memory: A New View of the Brain*. New York: Basic Books.

---. 1992. *The Strange, Familiar, and Forgotten: An Anatomy of Consciousness*. New York: Knopf.

Rubin, David C. 1995. *Memory in Oral Traditions: The Cognitive Psychology of Epic, Ballads, and Counting-out Rhymes*. New York: Oxford University Press.

Russell, Bertrand. 1943. *How to Read and Understand History: The Past as the Key to the Future*. Girard, Kansas: Haldeman-Julius.

Sacks, Oliver. 1985. *The Man Who Mistook His Wife for a Hat: And Other Clinical Tales*. New York: Summit Books.

---. 1989. *Seeing Voices: A Journey into the World of the Deaf*. Berkeley, California: University of California Press.

Schell, Jonathan. 1994. *Manchester Guardian*, London, 4 March 1994, 12.

Schmidt, Nathaniel. 1913. "Immortality" in *The New International Encyclopedia*. New York. X: 496.

Sereny, Gitta. 1995. *Albert Speer: His Battle with Truth*. New York: Knopf.

Shakespeare, William. 1992. *Hamlet*. Ed. Barbara A. Mowat and Paul Werstine. New York: Washington Square.

---. 1993. *A Mid-Summer Night's Dream*. Ed. Barbara A. Mowat and Paul Werstine. New York: Washington Square.

---. *Richard II*. 1996. New York: Washington Square..

Spence, Donald P. 1982. *Narrative Truth and Historical Truth: Meaning and Interpretation in Psychoanalysis*. New York: Norton.

Stoppard, Tom. 1993. *Arcadia*. Boston: Faber & Faber.

---. 1967. *Rosencrantz & Guildenstern Are Dead*. New York: Grove Press.

Strehlow, T. G. H. 1971. *Songs of Central Australia*. Sydney, Australia: Angus and Robertson.

Sutton, Peter, ed. 1988. *Dreamings: The Art of Aboriginal Australia*. New York: Penguin.

Thomas, Lewis. 1992. *The Fragile Species*. New York: Charles Scribner's Sons.

---. 1983. *Late Night Thoughts on Listening to Mahler's Ninth Symphony*. New York: Viking.

Virgil. 1983. *The Aeneid*. Trans. Robert Fitzgerald. New York: Random House.

Walsh, William S. 1893. *Handybook of Literary Curiousities*. Philadelphia: J. B. Lippincott.
White, Terence De Vere. 1957. *A Fretful Midge*. London.
Whitman, Walt. 1982. *Walt Whitman: Complete Poetry and Collected* Prose. Ed. Justin Kaplan. New York: Library of America, 1982.
---. 1982. "When Lilacs Last in the Dooryard Bloom'd." *Complete Poetry*. 459-67.
---. 1982. *Leaves of Grass*. 1855. *Complete Poetry*. 27-145.
Winson, Jonathan. 1985. *Brain and Psyche: The Biology of the Unconscious*. New York: Anchor/Doubleday.
Wordsworth, William. 1967. "The world is too much with us; late and soon." *Wordsworth: Poetry & Prose*. Ed. W. M. Merchant. Cambridge: Harvard University Press. 536.
Yates, Frances. 1966. *The Art of Memory*. Chicago: University of Chicago Press.

About the Author and Editor

Don Gifford was Professor of English and Class of 56 Professor of American Studies at Williams College, where he taught in a variety of disciplines including English Literature, American Studies, Environmental Studies, and the History of Ideas. He was a consultant on the psychology of invention with Arthur D. Little, Inc. and active in film making. Among his many publications were *Ulysses Annotated* (1988; original ed. 1974) that became a bestseller for the University of California Press, *Joyce Annotated: Notes for "Dubliners" and "A Portrait of the Artist as a Young Man"* (2nd ed 1982), *The Farther Shore: A Natural History of Perception, 1798-1984* (1990), and *The Literature of Architecture: The Evolution of Architectural Practice in 19th-Century America* (1966). The *New York Times*' reviewer became alternately "moved and fascinated" by Gifford's *The Farther Shore: A Natural History of Perception*.

Donald E. Morse, University Professor of American, Irish, and English Literature, University of Debrecen, and Emeritus Professor of English and Rhetoric, Oakland University, has been twice Senior Fulbright Professor (1987-89 and 1990-92) and twice Soros Professor at the University of Debrecen. Among his several books are *The Novels of Kurt Vonnegut: Imagining Being an American* (2003) and with Csilla Bertha *Worlds Visible and Invisible* (1994). In 2008 he and Bertha published their translations into English of *Silenced Voices: Hungarian Plays from Transylvania*—an undertaking supported by Rockefeller Study Fellowships. His more than 100 scholarly articles have appeared in journals throughout Europe, North America, Asia, and Brazil. In recognition of his international scholarship and service to Hungarian higher education he was awarded an honorary doctorate by the University of Debrecen and the László Országh Medal by the Hungarian Association for the Study of English.

Index

CPSIA information can be obtained at www.ICGtesting.com
Printed in the USA
BVOW041110141211

278300BV00002B/27/P

9 789042 032590